anatomy of
EXERCISE

PAT MANOCCHIA

FIREFLY BOOKS

A FIREFLY BOOK

Published by Firefly Books Ltd. 2008

Fourth printing, 2011

Publisher Cataloging-in-Publication Data (U.S.)

Manocchia, Pat.
    Anatomy of exercise : a trainer's inside guide to your workout / Pat Manocchia.
[192] p. : col. ill., photos. ;   cm.
Includes index.
Summary: With instructions and annotated anatomical illustrations, this book shows what happens to the body when one exercises. It is organized by body section covering core, legs and hips, back, chest, shoulders, arms, and outlines warm-up and stretching exercises.
ISBN-13: 978-1-55407-375-7 (bound)   ISBN-13: 978-1-55407-385-6 (pbk.)
ISBN-10: 1-55407-375-8 (bound)   ISBN-10: 1-55407-385-5 (pbk.)
1. Exercise — physiology.   2. Human mechanics.   3. Kinesiology.   I. Title.
612.76 dc22   QP303.M3663   2008

Library and Archives Canada Cataloguing in Publication

Manocchia, Pat
    Anatomy of exercise : a trainer's guide to your workout / Pat Manocchia.
Includes index.
ISBN-13: 978-1-55407-375-7 (bound)   ISBN-13: 978-1-55407-385-6 (pbk.)
ISBN-10: 1-55407-375-8 (bound)   ISBN-10: 1-55407-385-5 (pbk.)
1. Exercise. 2. Muscles — Anatomy.   3. Muscle strength.
4. Exercise — Physiological aspects.   I. Title.
GV461.M36 2008   613.7'1   C2007-905816-7

Published in the United States by
Firefly Books (U.S.) Inc.
P.O. Box 1338, Ellicott Station
Buffalo, New York 14205

Published in Canada by
Firefly Books Ltd.
66 Leek Crescent
Richmond Hill, Ontario L4B 1H1

For Hylas Publishing
Publisher: Sean Moore
Publishing Director: Karen Prince
Art Director: Gus Yoo
Editor: Amber Rose

Designers: Marian Purcell, Pleum Chenaphun, Lee Bartow, Terry Egusa
Anatomical Illustration: 3D4Medical
Photographer: Robert Wright
Models: Mark Tenore, Sydney Foster
Retoucher: Zoe Campagna

Printed in China

anatomy of
EXERCISE

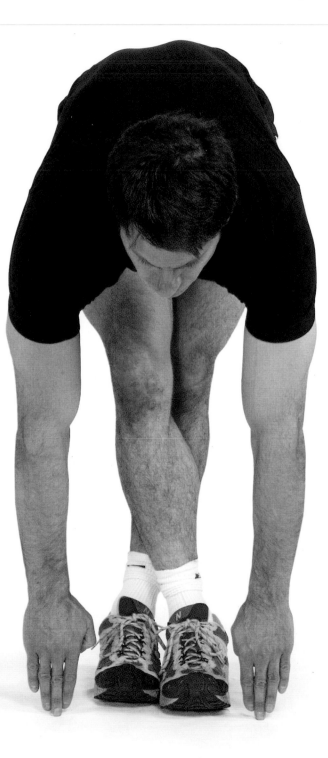

CONTENTS

INTRODUCTION

This is not the first book ever written that concerns itself with the anatomical structures that are involved in exercise, nor is it intended to be the final word. It is not meant to be an exhaustive exploration of exercise vocabulary either. This book takes a unique and comprehensive stance on the anatomy of exercise, useful to laymen and professional alike.

Predominately, books that have included exercises with anatomical representations, whether or not they included instructions for performing the exercises, were targeted at two groups of readers: body builders and scientists. What I've tried to do with this book is to make this kind of information accessible and useful to anyone who exercises, including bodybuilders and scientists. To that end, this book includes exercise types typically overlooked in similar works, such as aerobic activities, stretches, and stability work. Moreover, this book takes into consideration the ways in which the body's systems depend on one another to generate movement.

In other texts of this nature, exercises have for the most part been depicted as using a specific muscle group because the book was intended to show how to develop size or strength in that particular muscle or group. Unfortunately, what often doesn't get mentioned is how the adjacent muscles and structures, as well as some that are not directly or obviously involved, contribute to the exercises and subsequent improvement. Hip and spine position, for instance, contribute to almost every major exercise and are integral not only to the proper biomechanics of a given movement, but the subsequent improvement of the targeted muscle.

For each exercise, the muscles indicated in the illustrations are identified as the ones that are primarily involved in the movement, whether they are active or stabilizing. Active or primary muscles are defined as those that contract to move a structure, while stabilizing muscles are defined as those that either co-contract, or, by their activation, stabilize either the primary or a secondary structure to allow movement. In a push-up, for instance, the primary active muscles act to extend the elbow and adduct the humerus (upper arm) at the shoulder joint. Primary stabilizers act to ensure that the elbow and shoulder joints remain steady and track properly; however, without the contraction of the deep spinal and pelvic musculature, as well as anterior leg musculature that contract to keep those joints stable and allow the ankle joint to act as a fulcrum, the movement is not possible.

The contribution of the secondary stabilizers varies in degree, depending on the movement. For example, in a barbell curl, since the weight is in front of the body and is translated in a curvilinear fashion that creates a greater forward lever and subsequent need for stabilization as it moves upward, the back and hip muscles become more relevant with regard to movement contribution. If the movement could simply not be performed without the contribution of these muscles, they were included.

The point here is to make the reader aware that during any given movement, some muscles that may not play a major role in the actual execution may still be necessary contributors for proper biomechanics and form while the exercise is being performed. The basic method I used to determine this was to ask whether or not the movement could be performed if the secondary stabilizers were injured, but readers should be aware that the specifics are open to some debate.

There is an enormous amount of variation that can be made to these exercises, since for any one single exercise there are perhaps four or five different ways to alter the stimulus (by changing the grip, foot position, altering the speed of the movement, and so forth). I have included some of these variations for many of the exercises.

This book contains the basic exercise vocabulary that any program can, and should, be built around, whether you are an elite athlete, a raw beginner, or are suffering from an injury. The specific exercises to use as well as intensity (the weight used, when relevant), volume (number of sets and repetitions), duration (time per session), and frequency (sessions per week) will all be determined by your own specific capacities and goals. The best and most effective way to determine these things is to consult a professional in the fitness/wellness/strength training profession for a program and prescription that suits your unique abilities and objectives.

The text is laid out in a structure that mimics the progression of a typical workout. While the text encompasses all of the elements pursuant to a comprehensive workout, it is not intended to be prescriptive in any way. The best use of this book is as a reference manual for understanding both positioning and muscular involvement for the included exercises, and should stimulate some thought, when performing a given exercise, about how the rest of your body plays a part in any particular movement.

UPPER BODY (FRONT)

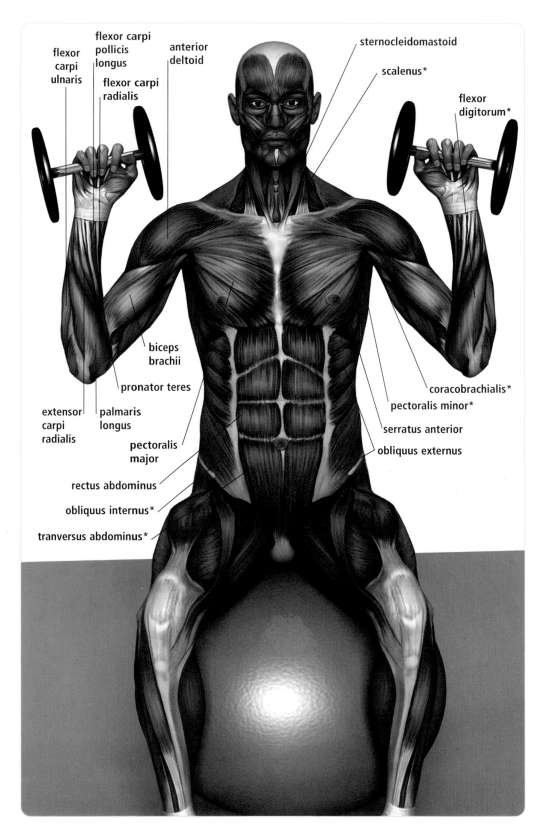

flexor carpi pollicis longus

flexor carpi ulnaris

flexor carpi radialis

anterior deltoid

sternocleidomastoid

scalenus*

flexor digitorum*

biceps brachii

pronator teres

coracobrachialis*

pectoralis minor*

extensor carpi radialis

palmaris longus

serratus anterior

obliquus externus

pectoralis major

rectus abdominus

obliquus internus*

tranversus abdominus*

UPPER BODY (BACK)

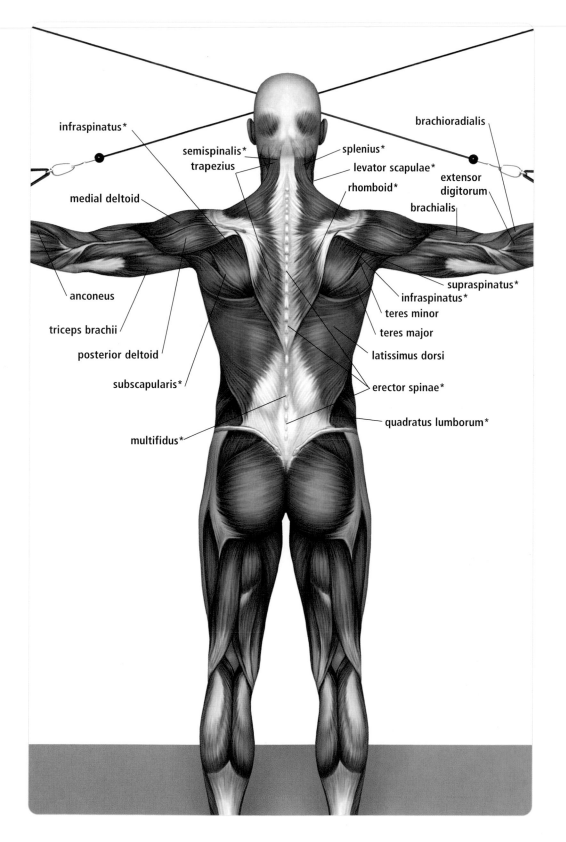

infraspinatus*

semispinalis*
trapezius

medial deltoid

anconeus

triceps brachii

posterior deltoid

subscapularis*

multifidus*

splenius*

levator scapulae*

rhomboid*

brachioradialis

extensor
digitorum

brachialis

supraspinatus*

infraspinatus*

teres minor

teres major

latissimus dorsi

erector spinae*

quadratus lumborum*

ANNOTATION KEY
* indicates deep muscles

LOWER BODY (FRONT)

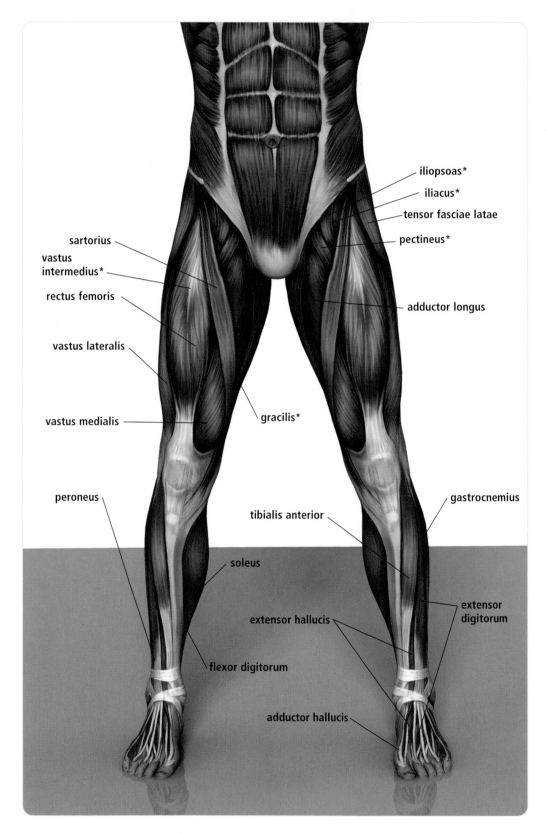

iliopsoas*

iliacus*

tensor fasciae latae

pectineus*

sartorius

vastus intermedius*

rectus femoris

adductor longus

vastus lateralis

vastus medialis

gracilis*

peroneus

tibialis anterior

gastrocnemius

soleus

extensor digitorum

extensor hallucis

flexor digitorum

adductor hallucis

ANNOTATION KEY

* indicates deep muscles

LOWER BODY (BACK)

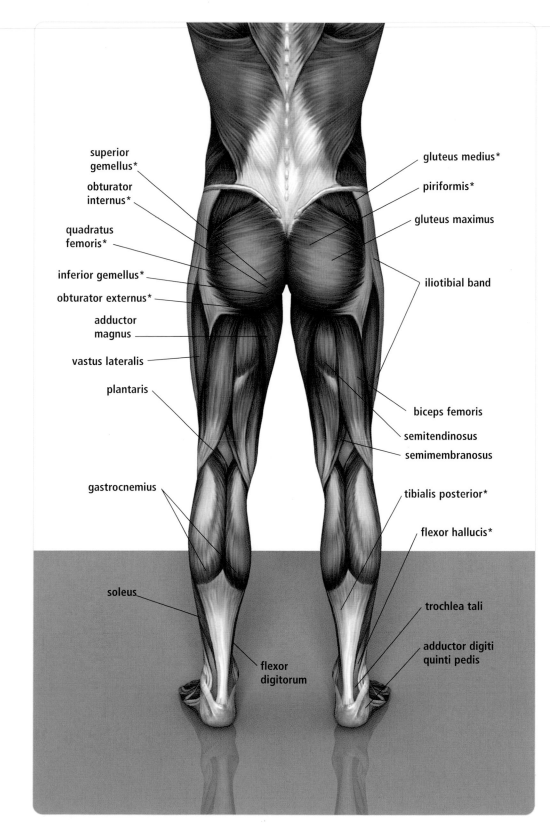

superior gemellus*

obturator internus*

quadratus femoris*

inferior gemellus*

obturator externus*

adductor magnus

vastus lateralis

plantaris

gastrocnemius

soleus

flexor digitorum

gluteus medius*

piriformis*

gluteus maximus

iliotibial band

biceps femoris

semitendinosus

semimembranosus

tibialis posterior*

flexor hallucis*

trochlea tali

adductor digiti quinti pedis

ANNOTATION KEY

* indicates deep muscles

WARM-UPS

The exercises included in the warm-up section require an increase in heart rate due to a demand for blood delivery. As a result, not only does oxygen and nutrient-rich blood become available to working muscles, but joint activity becomes easier, and body temperature due to this activity rises.

There are many effective and acceptable methods of increasing circulation and raising temperature both systemically (as depicted in these exercises) and locally in specific areas of the body. The one factor to keep in mind is that warm-ups should be performed ANY time exercise is done, in order to maximize benefit and minimize the potential for injury.

RUNNING

Starting Position: Stand with your knees slightly bent and your feet close together, directly under your hips. Your spine should be in a neutral position, with your head and chest up. Look directly forward.

Action: Correct form, often over-looked, is extremely important in order to derive the greatest benefit from running, and is described in detail here. Begin by rocking your body forward from your hips and torso. Allow your weight to roll forward onto the balls of your feet, allowing both heels to rise. Once your body begins to fall forward, remove one foot from the ground. As your body continues the forward motion, allow your forward foot to fall on the ball of the foot, not allowing your heel to hit the ground. Your foot should land directly under your center of mass while at the same time removing the supporting foot in the same manner.

Movement Path: Your center of mass (located in the region of your hips) is translated along a horizontal path directly forward.

The lower extremities move vertically, alternating in repetitive actions, directly beneath the center of mass.

LOOK FOR
- Each foot to strike the ground with the forward portion (your forefoot) directly beneath your center of mass
- An "S"-like shape when viewed in profile
- A soundless transition from foot to foot
- An even cadence of forefoot strikes

AVOID
- Reaching forward by extending your knee and landing in front of your center of mass
- Striking the ground with your heel, ever
- Vertical movement of your center of mass

STABILIZE BY
- Keeping your spine in a neutral position and your shoulders down and relaxed.
- Lifting your head and chin up.
- Keeping your arms bent and comfortable.

JUMP ROPE

Starting Position: Stand with your knees slightly bent and your feet close together, directly under your hips. Your spine should be in a neutral position, with your head and chest up. Look directly forward. With your elbows slightly bent and your palms facing forward, lightly grasp the handles of the jump rope keeping your elbows at your sides. The rope should be behind your heels.

Action: Bend your knees slightly and rotate your wrists backward and upward, bringing the rope up behind your body and over your head. As the rope begins the downward phase in front of your body, your wrists pull down and your palms rotate forward slightly, bringing the rope down force-fully toward the ground slightly in front of your toes. Jump just enough to allow the rope to pass beneath your feet and repeat.

Movement Path: There is a vertical translation of your entire body simulta-neously, while your wrists rotate in a forward direction.

STABILIZE BY
- Keeping your rib cage up and your spine in a neutral position
- Keeping your ankles, knees, and hips taut but relaxed

LOOK FOR
- Your ankles, knees, and hips to bend and extend with the same amplitude
- A bouncing movement
- A short foot-strike dura-tion, always landing on your forefeet
- The rope to have continuous tension
- Jumping to be relatively quiet

AVOID
- Allowing your heels to touch the ground
- Excessive arm movement
- Excessive knee or ankle movement
- Landing and pushing

BIKE

Starting Position: Sit on a bike, with your torso bent forward, resting your weight either on your hands or your elbows. Place your feet in the pedals with one leg flexed and bent up toward your torso and the other leg extended so that the pedal at the bottom range of the cycle allows your knee to extend fully, without locking it or requiring a hip shift.

Action: Extend one knee and hip, pushing down while simultaneously drawing the other knee and hip up. Repeat.

Movement Path: Your ankles and feet perform a repetitive alternating circular motion beneath and slightly forward of your center of mass.

Your knees and hips perform a pumping movement by flexing up and in toward the torso and then extending down and away.

STABILIZE BY
- Keeping your spine in a neutral position and your shoulders down

LOOK FOR
- An even movement between your two legs
- Your head and spine to remain stationary

AVOID
- Rounding your back or shoulders
- Focusing only on the extension of your knee and hip
- An improper seat height, i.e., either not allowing your knee and hip to fully extend, or requiring a change in your hip position to provide tension in the pedal through the bottom of the arc

ROWING

Starting Position: Sit in the rower so that your spine is vertical and your pelvis is tilted up. Your feet should be securely fastened. Bend your knees, sliding your torso forward, and grasp the handles.

①

STABILIZE BY
- Keeping your spine from head to tailbone rigid throughout the movement
- Maintaining a vertical alignment of your knees, ankles, and hips
- Keeping your shoulders down and away from your ears throughout the movement

Action: Push your feet into the foot stand by extending your knees, hips, and back. As your knees reach the midpoint of the range, your spine should be perpendicular to the ground and your elbows should be slightly bent as they pull the handle toward the midline of your torso. Finish the movement when your knees are completely extended, your spine is 5–10 degrees past perpendicular, and the handle is contacting your torso just below your chest, with your elbows retracted and your shoulders pulled down and back. Return by bending your knees and allowing your torso and arms to extend forward so that your torso is flexed forward, 5–10 degrees past perpendicular, and your arms are fully extended, so that your hands are above your ankles. Inhale on the downstroke and exhale as the stroke is executed.

LOOK FOR
- Your ankles, knees, hips, shoulders, and elbows to all move simultaneously
- Your forearms to remain horizontal at all times
- Your head and chest to remain high and forward

AVOID
- Rounding your back
- Excessive movement of your spine in either direction
- An unsynchronized movement pattern
- Allowing your knees to migrate either inward or outward

②

Movement Path: Your torso bends forward with your arms extended forward from the hips and your knees bent.

Your center of mass is translated horizontally as your torso flexes and extends along with your hips and knees.

STRETCHING

There are several different methods of stretching, only some of which are included here. The objective in performing these stretches is primarily to help improve range of motion of both the joints and muscles. They may be performed before exercising as a preparatory activity to stimulate neurological awareness, during an exercise session to provide blood to working muscles, after exercising as a method of cooling down and "reminding" joints and muscles of their movement patterns, or as a workout in themselves, as a method of recovery and regeneration from a prior bout of activity.

HIP-FLEXOR/HAMSTRING STRETCH

Starting Position: Kneel on one knee and place the opposite foot forward so that your front knee and hip are bent at 90 degrees. Form a direct line through your torso and spine to the knee on the floor. Let your hands rest by your sides and keep your shoulders down and relaxed and your chest up. Make sure your hips, knees, and ankles are in alignment.

Action: Raise your arms and move your hips forward simultaneously, until your arms are directly overhead and your front knee is over the toes of your front foot. Your torso remains vertical, and your bottom leg is opened at the hip joint. Continue by bringing your arms down and flexing your torso forward. Move your hips backward, straightening the front leg, while reaching your hands to the ground around your knee while your back knee remains on the ground.

LOOK FOR
- All joints to move at the same time

AVOID
- Allowing your front foot to leave the floor
- Any hip rotation
- Elevating your shoulders toward your ears

Movement Path: During the forward movement, your center of mass translates forward horizontally and slightly downward.
   During the backward movement, the same occurs in the opposite direction, with the exception of your torso flexing forward.

STABILIZE BY
- Keeping your hips parallel during all phases of the movement
- Keeping your front foot flat
- Maintaining the alignment of your hip, knee, and ankle throughout the movement

MODIFICATIONS
More Difficult: The forward movement is the same, but during the backward movement, let your back knee leave the ground and straighten your back leg as your hands reach directly down to the floor beneath your hips.

STRADDLE ABDUCTOR STRETCH

Starting Position: Spread your feet widely apart and bend your knees. Keep your hands on your knees and your weight evenly distributed throughout your feet. Your spine should be in a neutral position and your torso should bend forward at a 45 degree angle to the floor, so that your hips are behind your heels.

Action: Alternate leaning from side to side, bending the appropriate knee, while extending and straightening the opposite knee. Keep your hips back and your shoulders and chest facing forward.

Movement Path: Your hips remain behind your heels and your knees remain over your toes on the bent leg, as your center shifts from side to side, with a small arc in the movement (the highest point being the center of the arc and the lowest the widest).

STABILIZE BY
- Keeping your spine in a neutral position
- Keeping your feet flat and your weight evenly distributed
- Bearing some weight on your knees with your hands

LOOK FOR
- Your torso to remain stationary as it moves from side to side

AVOID
- Letting any part of your feet leave the floor
- Allowing your knees to extend beyond your toes
- Any rotation of your hips

CROSS LEG GLUTEAL STRETCH

Starting Position: Standing upright, cross your ankle to just above the opposite knee joint. Your lower body should appear as a figure 4. Let your arms hang at your sides (the position should be repeated on the opposite side after completeing the stretch).

STABILIZE BY
• Fixing and slightly arching your spine. If you are raising your arms to the front, stabilize your scapula and make sure your crossed leg is positioned firmly above your knee.

Action: Bend your standing leg by flexing your hip and knee to 90 degrees; essentially the stretch is a one-legged squat. Maintain the crossed leg in a stationary position. (To increase the intensity of the stretch apply a slight downward pressure to your crossed knee and a slight upward pressure on the same ankle). Hinge at your hips, not your spine.

LOOK FOR
• Your hips to move backward, prior to flexion at your knee
• Your lower leg to be perpendicular to the floor with your toes pointed forward
• A fixed spine and pelvic position with no anterior notation of pelvis
• Your head to remain in an upright position

AVOID
• Rounding your spine or tilting your pelvis
• Your front knee coming too far forward over your toes

Movement Path: To descend into the stretch position, your hips should move backward first. Lower until the femur of your standing leg is parallel to the floor (your knee should bend to 90 degrees). Raise your arms from your sides 90 degrees in front of you to your chest level, with your palms facing inward and your scapula stabilized.

PRESS-UPS

Starting Position: Lie flat on your stomach, with your palms down and next to your shoulders. Place your elbows by your sides, so that your forearms rest on the ground.

Action: Squeeze and depress your shoulder blades and squeeze your gluteals. Contract your spinal muscles and lift your chest and head upward and backward. As your chest rises, assist them by pressing your forearms and palms into the floor. Extend your elbows as far as your torso will allow without raising your hips off the floor, forcing your shoulders to shrug, or causing pain in your lower back or legs. Exhale as you rise and inhale as you return to the starting position.

LOOK FOR
- Your spine to curl on the way up
- Your arms to do no more than 50 percent of the work in lifting your torso

AVOID
- Elevating your hips off of the ground
- Forcing or pushing your torso upward with arms
- Raising or pinching your shoulders

Movement Path: Your spine moves in a curvilinear motion, both backward and upward, while your arms extend forward and downward.

STABILIZE BY
- Contracting your gluteals and pressing your hips to the floor
- Keeping your elbows at your sides
- Keeping your chest and head high

HIP/LOW BACK STRETCH

Starting Position: In a seated position with your torso vertical, extend one leg straight in front of you. Bend the other leg and cross it over the outstretched leg, so that the foot is flat on the floor and the ankle of the bent leg lies adjacent to the outside of the knee of the straight leg.

Action: Wrap the opposite arm around your bent leg. Pull your chest up and rotate your torso toward the bent knee by pulling around your leg. Reach directly behind your back with the hand on the bent–knee side, placing it on the ground and using it to keep your spine high and long while twisting. Hold the twisted position for 20 seconds. Repeat on the opposite side.

Movement Path: Your upper torso rotates with your spine vertical and perpendicular to your hips and the straightened leg, while your lower body remains stable.

LOOK FOR
- Your opposite arm to wrap around the bent knee with the crease in your elbow aligned just below your kneecap
- Your shoulders to remain in an even plane
- Your chest to stay high

AVOID
- Rounding your spine
- Letting any part of the foot on the bent-leg side rise off the ground

STABILIZE BY
- Using both arms to keep your torso in a vertical position
- Pushing the foot on the bent-knee side flat into the ground and keeping your hips stationary

QUAD STRETCH

Starting Position: Lie on your side. With your bottom hand, pull your bottom knee toward your chest. Grasp the shoelaces of your top leg, keeping your knees parallel.

Action: Contract your gluteals and extend your hip and heel backward. Gently assist the movement by pulling your upper leg back and upward, pausing at the farthest point in your range of motion. Return by flexing your hip joint and bringing the top knee forward until it is parallel with your bottom knee. Repeat on the opposite side.

LOOK FOR

- Your knee on top (the moving leg) to be pointing directly down from your hips at the bottom of the movement
- Your knee to be fully flexed during the movement
- Your leg and hip to be actively engaged during the entire process

Movement Path: Your torso and bottom leg are stationary as your upper leg and hip move from in front of your chest down and backward (your heel reaches toward your buttocks).

STABILIZE BY

- Pulling your abdomen in and keeping your hips parallel
- Keeping your chest and ribcage up and your shoulders back and down

AVOID

- Using your arm to generate the stretch (your arm should only assist with the movement, not create it)

ILLIO-TIBIAL BAND STRETCH

Starting Position: Standing up straight, cross one leg over the other, keeping your feet flat and slightly apart, with your back knee straight and your front knee only slightly bent.

Action: Keeping your legs crossed at the ankle, bend over completely to the ground while reaching your hands down toward your toes. Pause, then return to a standing position and uncross your legs. Cross your legs in the opposite direction and repeat the movement.

Movement Path: Your torso flexes directly forward.

LOOK FOR
• Your hands to remain fairly close to your body on the way down

AVOID
• Bouncing or forcing the extension of your hands toward the ground

STABILIZE BY
• Keeping your feet flat and your weight distributed slightly toward the back of the foot
• Keeping a slight bend in the front leg

UPPER BACK/SHOULDER STRETCH

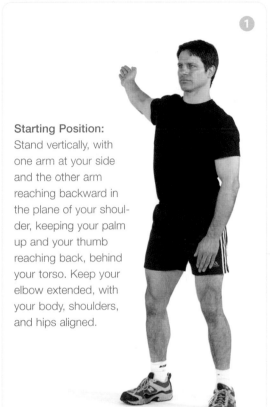

Starting Position: Stand vertically, with one arm at your side and the other arm reaching backward in the plane of your shoulder, keeping your palm up and your thumb reaching back, behind your torso. Keep your elbow extended, with your body, shoulders, and hips aligned.

STABILIZE BY
- Keeping your back straight, your chest up, and your hips facing forward

LOOK FOR
- Your shoulders to remain down throughout the entire movement
- Your head to face forward

AVOID
- Bending the elbow on the moving arm
- Letting the shoulder of the moving arm shrug upward
- Rotating any other part of your body

Action: Contract your chest and bring your arm forward in a horizontal plane. As your arm reaches the front of your body and is pointing forward, place the opposite hand and wrist under and around the moving arm and continue the movement by gently assisting the pressure of the arm across your chest. Repeat on the alternate side.

Movement Path: Your straight arm moves in a horizontal plane and a circular pattern around your body from front to back. Your legs, hips, and torso are stationary.

ACTIVE HAMSTRING STRETCH

Starting Position: Lie flat on your back with one knee bent and your foot flat on the floor. Clasp your fingers behind and below the knee on the opposite leg, with the knee joint on your upper leg pointing straight at the ceiling while keeping the leg relaxed.

①

LOOK FOR

- Your knee to be fully extended at the top of the movement (this may require changing your upper leg position to accommodate a full extension)
- An achievement of the end range of motion, or the farthest point at which the knee can straighten out

AVOID

- Flattening your lower back out during the knee extension
- Pulling on the extended leg with any appreciable force

Action: Extend the foot of the supported leg toward the ceiling by contracting your quadriceps (the front upper thigh), until your leg is fully extended. When fully extended, continue to contract your upper leg, bringing it up toward your chest. Use your hands to exert a VERY minimal, gentle pull upward in the same direction for no more than 2 seconds. Release the pull and allow your leg to drop slowly to its original position. Keep your foot flexed and your toes pulled up toward your shin.

Movement Path: Your lower leg moves in an upward arc toward the ceiling.

STABILIZE BY

- Keeping your spine in a neutral position
- Keeping the opposite foot flat on floor
- Making sure your hips remain stationary

②

ACTIVE HAMSTRING STRETCH

MODIFICATIONS
More Difficult: Keep one leg straight, bringing it up until you can grasp it behind the knee joint and gently assist it, while continuing to bring your leg toward your chest.

LEGS AND HIPS

The muscles of the legs span three joints, the hip, knee and ankle. The muscles of the upper leg operate the humerus at the hip and the tibia and fibula at the knee. Lower leg muscles move the foot. Multiple ligaments at each joint stabilize movement in all directions. The most critical function of the legs is interaction with hip, core, and back muscles to generate force, provide locomotion, and decelerate movement.

The muscles of the hips either originate or traverse the pelvis, and are responsible primarily for moving the legs. These muscles very often bear the brunt of improper training routines that tend to focus on movement in only one plane or direction and consequently develop strength imbalances that result both in poor performance and/or injury. In order for any force to be translated from the lower to upper body (particularly the spine), hip muscles must be addressed and trained properly.

LEG, HIP, BUTTOCKS ANATOMY

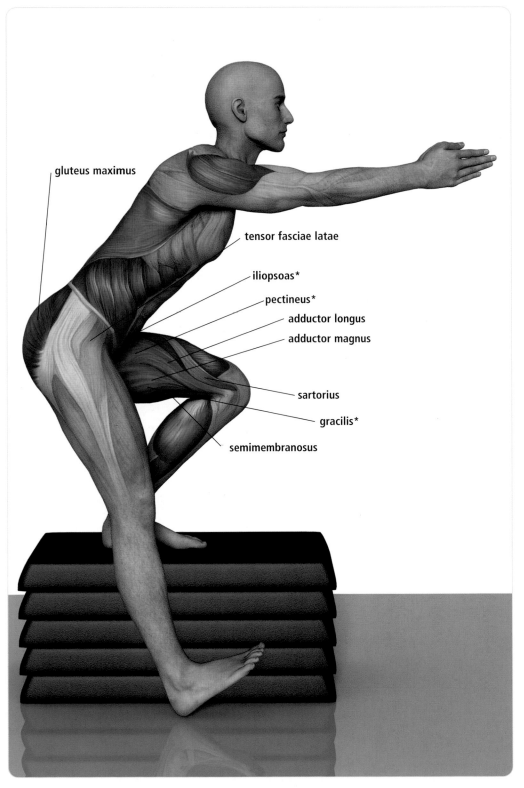

gluteus maximus

tensor fasciae latae

iliopsoas*

pectineus*

adductor longus

adductor magnus

sartorius

gracilis*

semimembranosus

ANNOTATION KEY

* indicates deep muscles

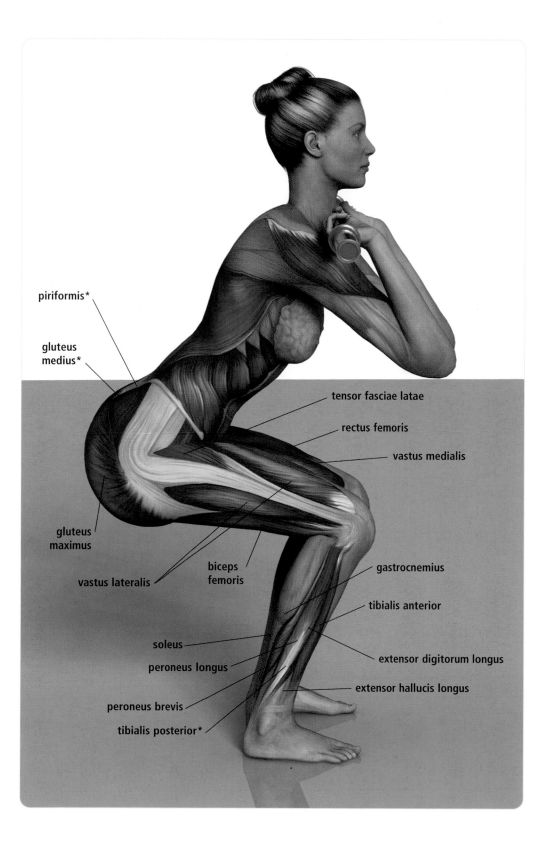

piriformis*

gluteus
medius*

tensor fasciae latae

rectus femoris

vastus medialis

gluteus
maximus

biceps
femoris

gastrocnemius

vastus lateralis

tibialis anterior

soleus

peroneus longus

extensor digitorum longus

extensor hallucis longus

peroneus brevis

tibialis posterior*

ANNOTATION KEY

* indicates deep muscles

BARBELL SQUAT

Starting Position: Stand with your feet slightly wider than shoulder width, resting a barbell across the top of your shoulder blades. Your hands should be wider than your hip width. Keep your chest up and your spine in a neutral position. Keep your eyes looking up, 20 degrees above horizontal. Your weight should be spread evenly across your feet.

Action: Retract your hips, pulling them backward. Bend your knees, keeping your spine in a neutral position, allowing your body to hinge at the hip and knee joints, until your upper legs are parallel with the floor and your spine is at a 45 degree angle to the floor. Your knees should extend slightly forward, with your chest and head up. Your knees should be directly over the front of your feet.

LOOK FOR
- The barbell to drop down in a vertical line, directly above the mid-foot to heel
- All joints to move at the same time
- Balance throughout the movement

AVOID
- Sliding your knees forward, beyond your toes
- Rounding your back
- Allowing the bar to roll up on your neck, toward your head
- Letting your knees slide until they are either wider or narrower than your feet

Movement Path: Your hips and spine slide down and backward while your knees slide forward until the bottom of the movement; then your knees, hips, and back move straight up simultaneously.

STABILIZE BY
- Keeping your spine in a neutral position and your shoulder blades pinched down and back
- Keeping your abdominal muscles taut and your hands on the bar, with some tension in your grip

BARBELL SQUAT

1

- trapezius
- posterior deltoid
- infraspinatus*
- quadratus lumborum*
- erector spinae*
- transversus abdominis*
- obliquus externus
- iliopsoas*
- tensor fascia latae
- peroneals
- tibialis anterior

2

- vastus intermedius
- rectus femoris
- vastus medialis
- gluteus medius*
- gluteus maximus
- biceps femoris
- vastus lateralis
- soleus
- gastrocnemius

BEST FOR

- biceps femoris
- gluteus maximus
- gluteus medius
- rectus femoris
- vastus intermedius
- vastus lateralis
- vastus medialis

ANNOTATION KEY

Black text indicates active muscles

Gray text indicates stabilizing muscles

* indicates deep muscles

FRONT SQUAT

Starting Position: Stand with your chest and head up, your arms extended forward, and a barbell resting on the top of your shoulder muscles. Keep your spine neutral and your feet slightly wider than shoulder width.

Action: Retract your hips and bend your knees, keeping your head and chest up and your spine in a neutral position. Lower your hips until your upper legs and thighs are at least parallel with the ground. Your head and spine should be at a 45 degree angle to your hips and to the ground. Your hands can either be extended forward or crossed to grip the bar so that it does not slip.

STABILIZE BY
- Retracting your shoulder blades, keeping your abdomen pulled up and in, and keeping your knees parallel and over your feet

LOOK FOR
- A slow, controlled descent and ascent
- No movement in the vertical line of the bar as it drops
- No foot movement or heel lift

AVOID
- Extending your knees beyond your toes
- Dropping your elbows; keep them and your upper arms parallel to the ground at all times
- Extending your head forward or elevating your scapula and shoulders

Movement Path: Drop straight down and push straight up by extending your arms on the way down and extending your legs on the way up. Pushing through your knees, your hips, shoulders, and ankles move up simultaneously.

FRONT SQUAT

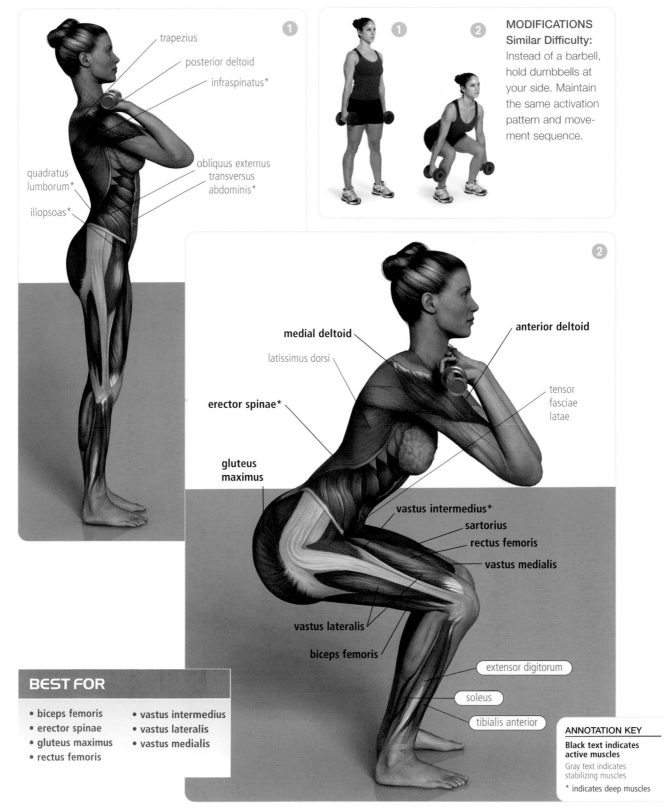

trapezius

posterior deltoid

infraspinatus*

quadratus lumborum*

obliquus externus

transversus abdominis*

iliopsoas*

MODIFICATIONS
Similar Difficulty:
Instead of a barbell, hold dumbbells at your side. Maintain the same activation pattern and movement sequence.

medial deltoid

anterior deltoid

latissimus dorsi

tensor fasciae latae

erector spinae*

gluteus maximus

vastus intermedius*

sartorius

rectus femoris

vastus medialis

vastus lateralis

biceps femoris

extensor digitorum

soleus

tibialis anterior

BEST FOR

- biceps femoris
- erector spinae
- gluteus maximus
- rectus femoris
- vastus intermedius
- vastus lateralis
- vastus medialis

ANNOTATION KEY

Black text indicates active muscles
Gray text indicates stabilizing muscles
* indicates deep muscles

LUNGE

Starting Position: Stand with your feet close together and your hands on your hips.

Action: Keeping your head up, your spine in a neutral position, and your hands on your hips, take a step forward, bending your front knee to a 90 degree angle and dropping your front thigh until it is parallel with the ground. Your back knee drops straight down behind you, so that you are balancing on the toe of your foot to create a 90 degree angle in your knee joint and a straight line from your spine through your bottom knee. Return to the starting position by pushing on your front foot and elevating with your back leg until standing.

Movement Path: The general motion is forward and descending. Your spine stays in a vertical position and is translated forward and downward by the step and the descent.

LOOK FOR
- No translation forward from the hips (do not bend)
- Your spine to remain in the same position as it moves down and up
- No lateral movement of your leg as you step, either landing or pushing

AVOID
- Raising the heel of your stepping foot off the ground or rotating your hips or torso

STABILIZE BY
- Keeping your chest high, stomach up, and spine neutral
- Evenly distributing your weight across your front foot, from front to back
- Keeping your back foot on the toe and your weight in the back of the stepping leg

LUNGE

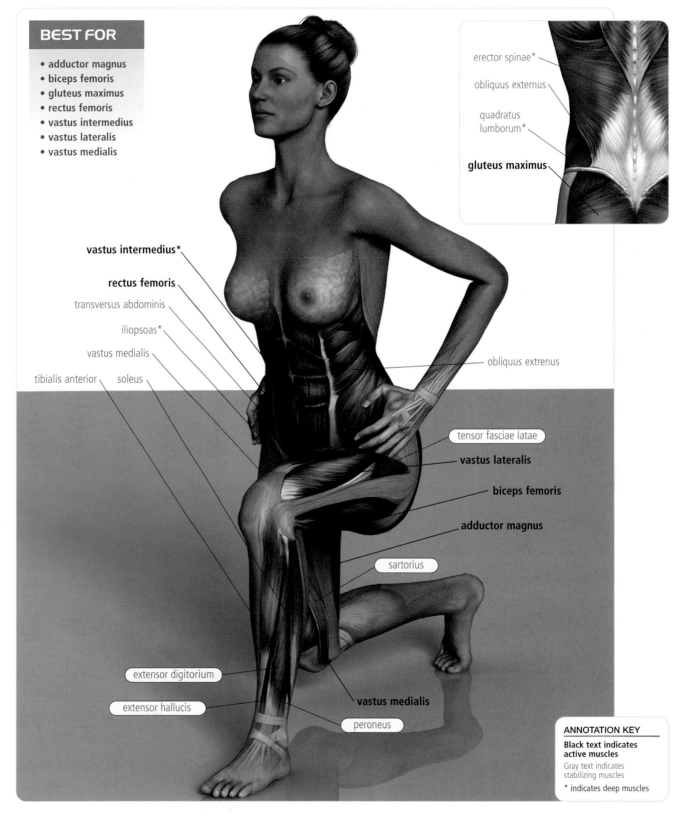

BEST FOR

- adductor magnus
- biceps femoris
- gluteus maximus
- rectus femoris
- vastus intermedius
- vastus lateralis
- vastus medialis

erector spinae*

obliquus externus

quadratus lumborum*

gluteus maximus

vastus intermedius*

rectus femoris

transversus abdominis

iliopsoas*

vastus medialis

tibialis anterior

soleus

obliquus extrenus

tensor fasciae latae

vastus lateralis

biceps femoris

adductor magnus

sartorius

extensor digitorium

extensor hallucis

vastus medialis

peroneus

ANNOTATION KEY

Black text indicates active muscles

Gray text indicates stabilizing muscles

* indicates deep muscles

LATERAL LUNGE

Starting Position: Stand vertically with your feet directly below your hips and your hands on your hips.

LOOK FOR

- A simultaneous movement of your arms and hips
- Your chest to remain up and your shoulders to remain down

AVOID

- Any part of the stepping foot leaving contact with the ground or your knee extending forward beyond your toe
- An excessive drop in torso angle beyond or below 45 degrees

Action: Step directly out to the side at 180 degrees, retracting your hips and keeping your spine neutral. As your chest moves forward and your hips retract, extend your arms in order to ensure balance. Stop at the bottom of the movement when the upper thigh of your stepping leg is parallel to the ground. The opposite knee should be extended, your hips should be behind the stepping foot, and your knee should not exceed the toe line and should be directly over the foot. Your upper arms should be parallel with the ground. Pushing back off the stepping leg, return to the starting position.

Movement Path: You move laterally to the side; your arms go forward and your hips go back.

Your torso drops as your hips retract.

Use one foot as both the decelerator and accelerator. Use the standing or stationary foot as a balance lever.

STABILIZE BY

- Keeping your hips retracted, your chest up, and using your arms as a counterbalance to the retraction of your hips
- Keeping the opposite leg in contact with the floor, and maintain tension on your quadriceps and hamstrings, so that your knee is locked and extended

LATERAL LUNGE

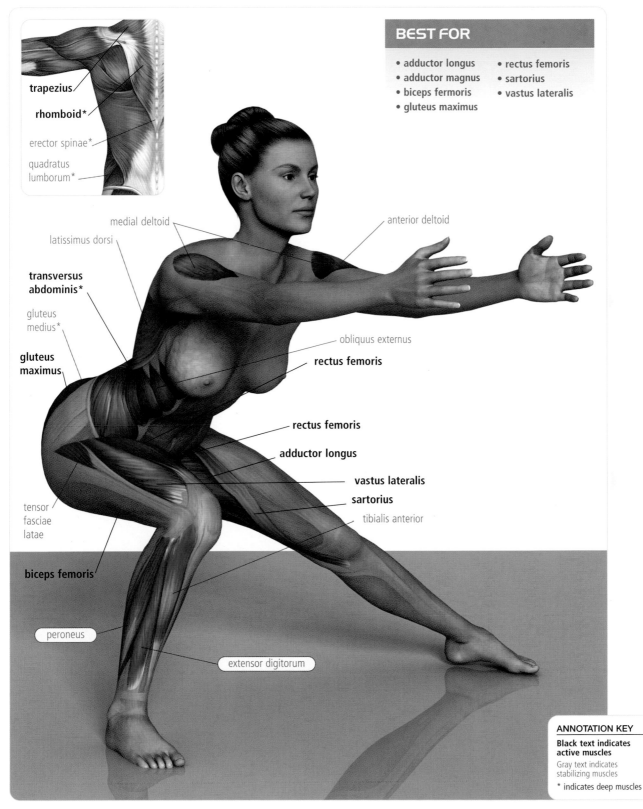

trapezius

rhomboid*

erector spinae*

quadratus lumborum*

BEST FOR

- adductor longus
- adductor magnus
- biceps fermoris
- gluteus maximus
- rectus femoris
- sartorius
- vastus lateralis

medial deltoid

anterior deltoid

latissimus dorsi

transversus abdominis*

gluteus medius*

gluteus maximus

obliquus externus

rectus femoris

rectus femoris

adductor longus

vastus lateralis

sartorius

tibialis anterior

tensor fasciae latae

biceps femoris

peroneus

extensor digitorum

ANNOTATION KEY

Black text indicates active muscles

Gray text indicates stabilizing muscles

* indicates deep muscles

STEP-UP

1 **Starting Position:** Place one foot on a block in front of you, preferably with the raised knee at close to a 90 degree angle. Make sure your body is vertical, your chest is up, and your knee is directly over your foot. Your raised knee should not exceed the toe line and your foot should be flat on the surface of the step. Grasp the dumbbells by the side of each hip.

2

LOOK FOR
- A slight forward translation and directly upward movement of your spine

AVOID
- Bending or extending your back knee
- Allowing your front knee to slip forward beyond the toe line or any part of your front foot to lift off the step
- Moving your knee either laterally or medially; keep it directly over the stepping foot

Action: Begin by slightly leaning forward. Allow the weights to swing forward so that they are close to the plane of your ankle. Keeping your back leg straight, push through your top leg, extending your knee and hips simultaneously to drive your body up and over the step. Do not allow your back leg to push off the floor.

Movement Path: The movement is slightly forward and directly upward.

Your head should begin behind the elevated foot and end directly over it.

Allow your arms to simply stabilize the weight and hang as pendulums, following the body's natural path.

Descend in the same fashion.

STABILIZE BY
- Keeping your upper back muscles and shoulders down and back
- Not allowing your momentum to bring forward either the weights or your torso
- Keeping your hip, shoulder, and ankle in a line from the bottom weight

STEP-UP

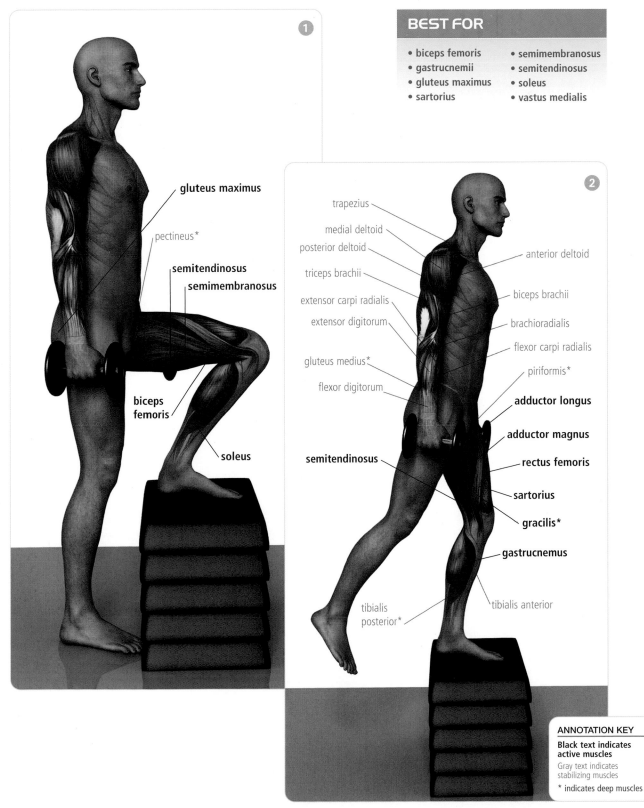

1

gluteus maximus

pectineus*

semitendinosus

semimembranosus

biceps femoris

soleus

BEST FOR

- biceps femoris
- gastrucnemii
- gluteus maximus
- sartorius
- semimembranosus
- semitendinosus
- soleus
- vastus medialis

2

trapezius

medial deltoid

posterior deltoid

triceps brachii

extensor carpi radialis

extensor digitorum

gluteus medius*

flexor digitorum

semitendinosus

anterior deltoid

biceps brachii

brachioradialis

flexor carpi radialis

piriformis*

adductor longus

adductor magnus

rectus femoris

sartorius

gracilis*

gastrucnemus

tibialis anterior

tibialis posterior*

ANNOTATION KEY

Black text indicates active muscles

Gray text indicates stabilizing muscles

* indicates deep muscles

STEP-DOWN

Starting Position: Begin by standing in a vertical position on a block, with one foot unsupported off to the side.

STABILIZE BY
• Keeping your spinal muscles active, your shoulders retracted and depressed, the opposite leg extended and taut, and your opposite foot flexed

Action: Drop your torso by retracting the hip over the standing foot and bending your knee, allowing your torso to ride forward. Extend your arms, dropping the non-weight bearing leg to below the level of the step. Drop to just above 90 degrees, or to a point at which balance can be achieved. Return by extending your hips, straightening your knee, and lifting your chest up and away.

LOOK FOR
• Your head to remain directly above your ankle
• Your knee and hips to move simultaneously

AVOID
• Extending your knee beyond your toe line, any rotation in your hips or torso, and any deviation of the standing knee from above the weight-bearing foot

Movement Path: While descending directly down, your arms extend and your hips retract. Use your hands and arms for balance.

STEP-DOWN

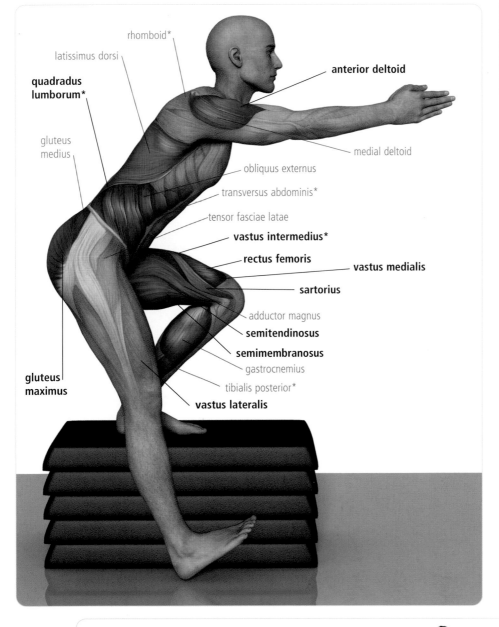

rhomboid*

latissimus dorsi

quadradus lumborum*

gluteus medius

gluteus maximus

anterior deltoid

medial deltoid

obliquus externus

transversus abdominis*

tensor fasciae latae

vastus intermedius*

rectus femoris

vastus medialis

sartorius

adductor magnus

semitendinosus

semimembranosus

gastrocnemius

tibialis posterior*

vastus lateralis

BEST FOR

- gluteus maximus
- rectus femoris
- semimembranosus
- semitendinosus
- vastus intermedius
- vastus lateralis
- vastus medialis

ANNOTATION KEY

Black text indicates active muscles

Gray text indicates stabilizing muscles

* indicates deep muscles

MODIFICATIONS

More Difficult: Grasp a weight, such as a medicine ball, at your chest in the initial standing position. As you descend by retracting your hips and bending your knee, keeping your spine taut and your chest up, extend the weight forward so that at the bottom of the movement the weight and your arms are fully extended from your torso and balance is achieved. Return to the starting position by extending your hip and knee and elevating your torso, while at the same time drawing the weight back toward your chest to the starting position.

CALF RAISE

Starting Position: With your body vertical, place the front part of one or both feet (from the ball joint to the toe) on a step. Either one or both feet should be extended, so that the heel and arch of each foot is beyond the edge of the step. Your hips, ankle(s), and shoulders should be aligned; your spine should be in a neutral position and your head should be up. Grasp a dumbbell in one hand if using only a single leg.

Action: Allow your heels to descend in a vertical line, dropping toward the floor. At the bottom of the motion, push down on your toes and the arch of the foot, keeping your spine straight and elevating your body directly upward.

Movement Path: Your body moves in a vertical descent and ascent. Use the front part of your foot as the activator.

LOOK FOR
- Both heels to drop at the same time

AVOID
- Leaning forward or lurching your torso
- Bending your knees or hips
- Any spinal movement at all

STABILIZE BY
- Keeping your spine in a neutral position
- Keeping your feet evenly placed on the step and your upper legs locked
- Placing the fingertips of one or both hands on supportive surfaces. Do not allow your hands to influence movement up or down; simply allow them to help stabilize your spinal position and torso angle

CALF RAISE

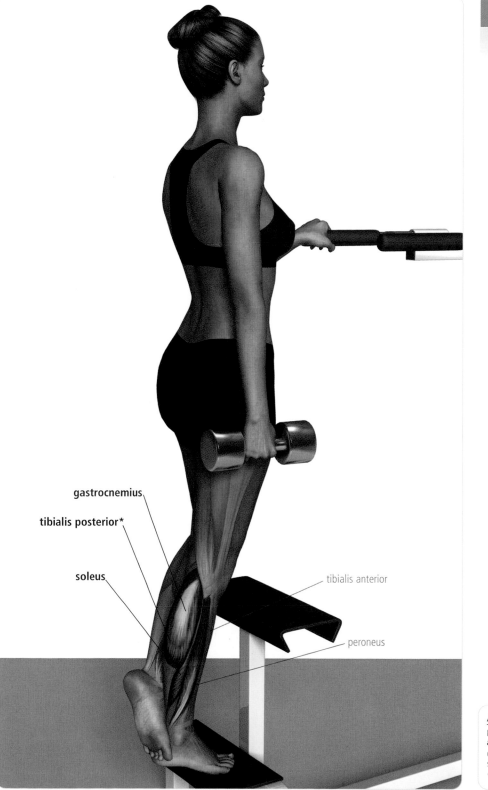

gastrocnemius

tibialis posterior*

soleus

tibialis anterior

peroneus

BEST FOR

- gastrocnemii
- soleus
- tibialis posterior

ANNOTATION KEY

Black text indicates active muscles

Gray text indicates stabilizing muscles

* indicates deep muscles

CLAMSHELLS

Starting Position: Lie on your side with your knees bent so that your shoulders, hips, and ankles are in a straight line. Make sure your spine is in a neutral position and that your bottom arm is bent with your head resting on it to allow it to remain in line with your spine. Hold a weight firmly against your thigh.

1

Action: Raise your upper knee by rotating your hip toward the ceiling, contracting your gluteals. The toes on your upper foot rise too as the ankle on the active leg remains in a fixed position relative to the leg. Return to the starting position by lowering the knee joint.

Movement Path: Both your upper and lower leg move in a curvilinear path away from the body's midline, from horizontal toward vertical and back.

2

LOOK FOR
• The only movement to happen below the crease in the hip joint
• The active leg to appear as if it were on a hinge

AVOID
• Any spinal movement
• Any pelvic movement

STABILIZE BY
• Keeping your stomach muscles pulled in and up throughout the entire movement
• Maintaining a neutral spinal position throughout the movement
• Keeping the ankle on the active leg locked in a fixed position
• Firmly controlling the hip joint on the active side

CLAMSHELLS

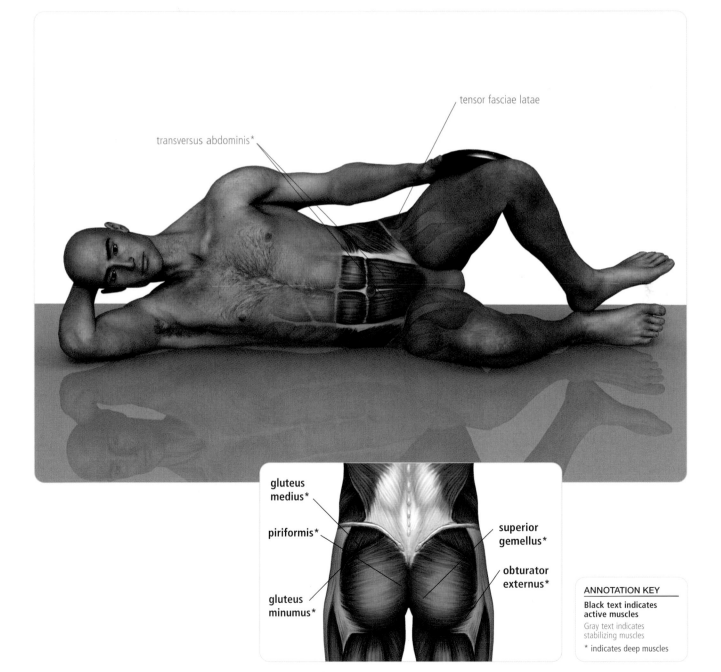

transversus abdominis*

tensor fasciae latae

gluteus
medius*

piriformis*

gluteus
minumus*

superior
gemellus*

obturator
externus*

ANNOTATION KEY

**Black text indicates
active muscles**

Gray text indicates
stabilizing muscles

* indicates deep muscles

LEG PRESS PLIÉ

①

Starting Position: Sit with your feet placed slightly wider than shoulder width. Ensure that they are flat, directly under your hips. Grasp the handles firmly, keeping your chest and ribcage high. Make sure your knees are directly above your toes and pointing in the same direction.

Action: Using the entirety of each foot, push firmly into your foot base, extending both knees and hips simultaneously while exhaling until your knees are fully extended. Inhale, allow your knees, hips, and feet to return to the starting position.

LOOK FOR
- A simultaneous extension of both your knee and hip joints as well as a slight flexion of your ankle joints
- Your gluteals to remain in firm contact with the seat

AVOID
- Allowing your knees to migrate either toward or away from each other
- Rotating your toes outward
- Allowing your hips or gluteals to roll up off of the seat

②

Movement Path: Your torso remains in a constant position as your legs and hips are extended, so that your center of mass moves away from your feet in a straight line horizontally. The return path is the same but opposite action.

STABILIZE BY
- Pulling your hips down firmly into the seat
- Keeping your feet flat
- Keeping your knees parallel and over your feet at all times

LEG PRESS PLIÉ

BEST FOR

- adductor longus
- adductor magnus
- gluteus maximus
- gluteus medius
- gracilis
- rectus femoris
- sartorius
- semimembranosus
- semitendinosus
- soleus
- vastus medialis

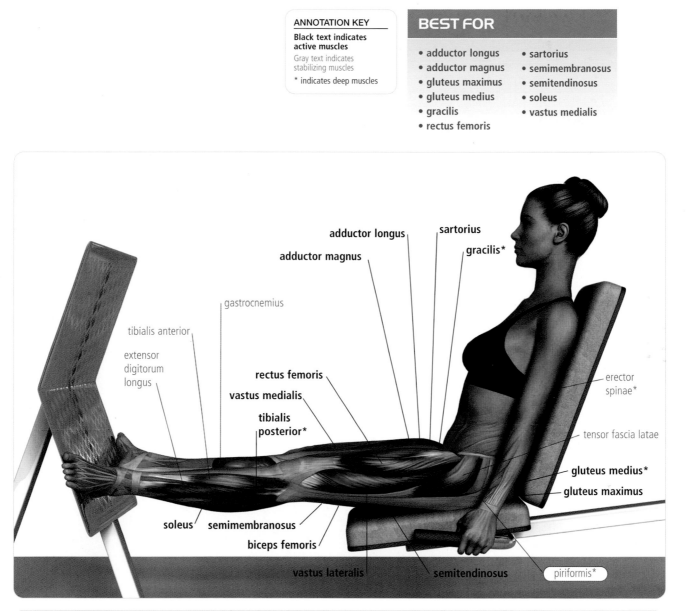

adductor longus
adductor magnus
sartorius
gracilis*
gastrocnemius
tibialis anterior
extensor digitorum longus
rectus femoris
vastus medialis
tibialis posterior*
erector spinae*
tensor fascia latae
gluteus medius*
gluteus maximus
soleus
semimembranosus
biceps femoris
vastus lateralis
semitendinosus
piriformis*

MODIFICATIONS

More Difficult: Use only one leg, starting with the active leg bent at a 90 degree angle and your foot at chest height. Keep your opposite leg straight and avoid rotating your hips or pelvis. Maintain the same activation pattern and movement sequence.

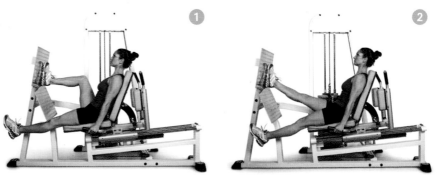

LEG EXTENSION

Starting Position: In a seated position, ensure that the backs of your knees rest directly against the edge of the seat. Your knee joints should be directly adjacent and bent to at least 90 degrees to the machine's pivot arm. The resistance pad should lie just above ankle height. Grasp the handles firmly, with your torso upright and your hips pulled down securely into the seat.

LOOK FOR
- Only your lower legs to move; the entire rest of your body should remain stationary
- Control of the movement in both directions

AVOID
- Allowing any space between the back of your knees and the seat edge
- Any spinal movement whatsoever
- Elevating your shoulders
- Allowing your hips or gluteals to elevate off of the seat

Action: Exhale while contracting your quadriceps and extending your legs, kicking your feet outward and up until your knee joints are completely extended. Allow your lower legs to drop while inhaling to return to the starting position.

Movement Path: Your feet are kicked out and away from under the upper leg though a curvilinear arc.

STABILIZE BY
- Keeping your hips firmly down into the seat throughout the movement
- Maintaining an upright torso, right against the seat back
- Pulling your toes up slightly so that your ankle joints are slightly dorsi-flexed

LEG EXTENSION

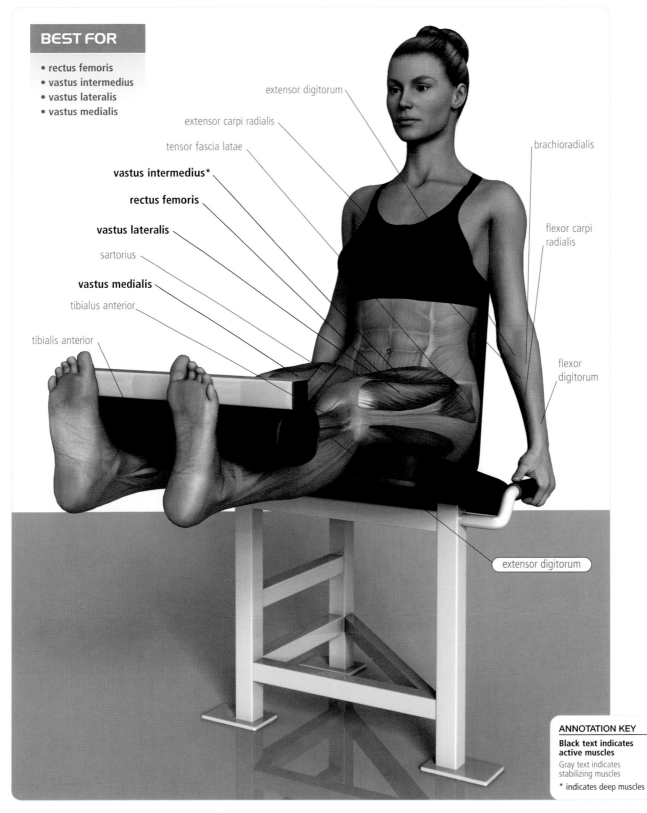

BEST FOR

- rectus femoris
- vastus intermedius
- vastus lateralis
- vastus medialis

extensor digitorum

extensor carpi radialis

tensor fascia latae

vastus intermedius*

rectus femoris

vastus lateralis

sartorius

vastus medialis

tibialus anterior

tibialis anterior

brachioradialis

flexor carpi radialis

flexor digitorum

extensor digitorum

ANNOTATION KEY

Black text indicates active muscles

Gray text indicates stabilizing muscles

* indicates deep muscles

LEG CURL

Starting Position: Lie in a prone position, with your hips directly over the apex of the bend in the bench. Adjust your lower legs so that the resistance arm pad rests on your achilles tendon, just above the top of your shoe. Your knee joint should be directly adjacent to the pivot joint of the machine.

Action: Exhaling, contract your hamstrings and pull your heels up and in until the bar comes into contact with your lower gluteals (maximum range) or until your knee joint reaches 90 degrees of flexion. Inhale as the weight is lowered in a controlled movement to the starting position.

Movement Path: From a 180 degree flat plane with your knees completely extended, your lower legs swing up in an arc toward your hips until there is less than 90 degrees of knee flexion and your ankles are close to your hips. Return in the same curvilinear path.

STABILIZE BY
- Keeping your hips in contact with the bench at all times
- Firmly grasping the handles
- Keeping your feet and ankles at a 90 degree angle

LOOK FOR
- The only movement to occur from the knee joint to the ankle joint as your lower leg is drawn up toward your hips

AVOID
- Parting your toes
- Arching your back
- Allowing your hips to elevate off of the bench
- Letting the resistance pad slide on the lower leg as your muscles are contracted and your knee is flexed

LEG CURL

BEST FOR

- biceps femoris
- gastrocnemius
- semimembranosus
- semitendinosus

biceps femoris

flexor digitorum

gastrocnemius

gluteus medius*

semimembranosus

semitendinosus

obliquus externus

latissimus dorsi

gluteus maximus

biceps brachii

soleus

extensor hallucis

flexor digitorum

tibialis anterior

extensor digitorum

extensor digitorum longus

brachioradialis

extensor carpi radialis

flexor carpi radialis

ANNOTATION KEY

Black text indicates active muscles

Gray text indicates stabilizing muscles

* indicates deep muscles

WALL SIT

Starting Position: Place your back against a wall in a standing position, with your feet directly under your hips and your hands placed against the wall for balance. Walk your feet forward and slowly slide your torso down the wall until your hips, knees, and ankles are all at 90 degree angles. Your knees should be directly over your ankles. Your legs should be parallel. Make sure your feet are flat on the ground. Keep your spine in a neutral position, your hands on your hips, and your head up.

Action: None (isometric muscle contraction). The position should be maintained until fatigue makes your muscles burn and the form is difficult to maintain.

Movement Path: None.

LOOK FOR
- No movement whatsoever
- 90 degree angles
- Your muscles to shake when fatigued

AVOID
- Any movement
- Reaching a level of fatigue which requires you to sit rather than stand to remove yourself from the wall

STABILIZE BY
- Keeping your shoulders, hips, and upper back pressed firmly into the wall
- Keeping your feet flat and the pressure evenly distributed
- Maintaining tension on the transverse abdominis by pulling it up and in with your ribs high

WALL SIT

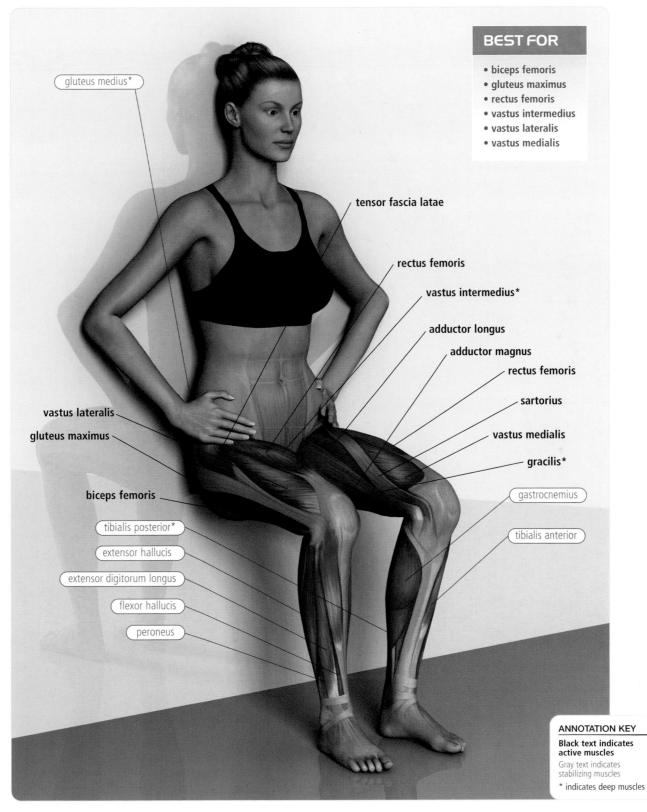

gluteus medius*

BEST FOR

- biceps femoris
- gluteus maximus
- rectus femoris
- vastus intermedius
- vastus lateralis
- vastus medialis

tensor fascia latae

rectus femoris

vastus intermedius*

adductor longus

adductor magnus

rectus femoris

sartorius

vastus medialis

gracilis*

gastrocnemius

tibialis anterior

vastus lateralis

gluteus maximus

biceps femoris

tibialis posterior*

extensor hallucis

extensor digitorum longus

flexor hallucis

peroneus

ANNOTATION KEY

**Black text indicates
active muscles**

Gray text indicates
stabilizing muscles

* indicates deep muscles

THE SKATER

Starting Position: Bend your knees and hips so that your torso is at a 45 degree angle. Place your feet slightly closer than hip width (the active side hip should be flexed to a slightly greater degree). Hold the cable frame in front of you with your hands up. Standing on one foot, attach the other by an ankle strap to a low pulley located directly in front of the involved leg. Raise the involved leg slightly off of the ground.

STABILIZE BY
- Keeping your spine in a neutral position
- Keeping the standing leg in a fixed position and your weight evenly distributed

LOOK FOR
- The sole of your foot to rotate outward at the same time as the knee joint
- The movement to appear balanced and timed so that it flows easily

AVOID
- Rotating your upper body
- Excessively arching your back
- Gripping the column so that balance is transferred from your standing foot to your hands

Action: Your knee extends as the active hip both extends and externally rotates, until your hip and knee are fully extended. Return by allowing the resistance to pull your leg back to the starting position.

Movement Path: Your foot moves away and out from both your midline and center of mass on the extension and up and into them on return to the starting position.

THE SKATER

BEST FOR

- gluteus maximus
- gluteus medius
- gluteus minimus
- obliquus externus
- obturator externus
- piriformis
- quadratus lumborum
- tensor fasciae latae
- vastus lateralis
- vastus medialis

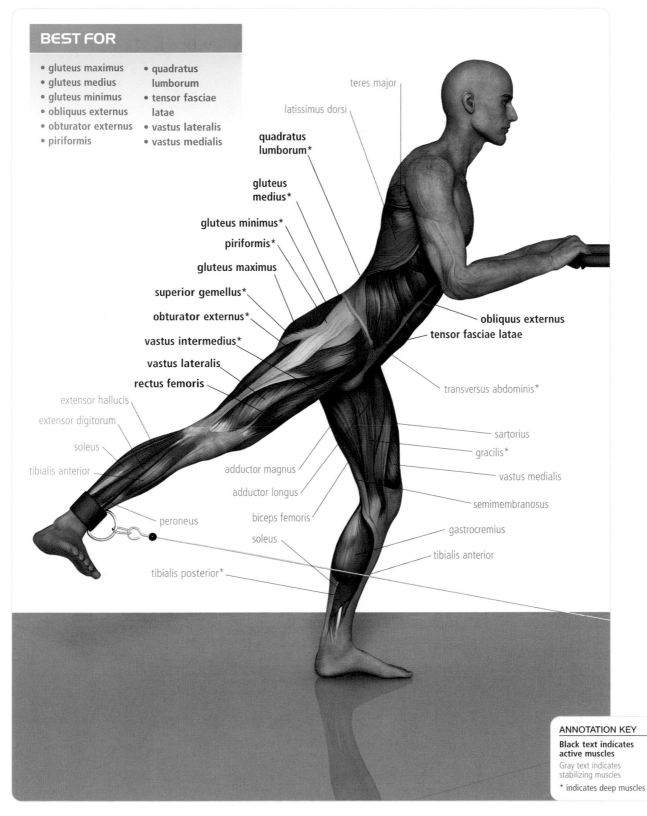

teres major

latissimus dorsi

quadratus lumborum*

gluteus medius*

gluteus minimus*

piriformis*

gluteus maximus

superior gemellus*

obturator externus*

vastus intermedius*

vastus lateralis

rectus femoris

extensor hallucis

extensor digitorum

soleus

tibialis anterior

peroneus

adductor magnus

adductor longus

biceps femoris

tibialis posterior*

soleus

obliquus externus

tensor fasciae latae

transversus abdominis*

sartorius

gracilis*

vastus medialis

semimembranosus

gastrocremius

tibialis anterior

ANNOTATION KEY

Black text indicates active muscles

Gray text indicates stabilizing muscles

* indicates deep muscles

PLOUGH

Starting Position: Place your hands on the ground, with your legs extended so that the tops of your shoes are on top of a Swiss ball in a push-up position. Keep your spine neutral.

LOOK FOR
• A simultaneous movement while your hips raise, so that your spine is at a 45 degree angle from your hip to your shoulder from the ground

AVOID
• Dropping your knees toward the floor
• Bending your elbows
• Allowing your shoulders to either elevate toward your ears or round forward

Action: Pull your knees up toward your chest while flexing your feet, balancing your toes on the ball, driving your hips toward the ceiling and retracting your abdomen.

Movement Path: Your torso flexes in a straight line and a single plane.
   Your feet move up toward your midline in a horizontal plane.

STABILIZE BY
• Keeping your chest high and contracted
• Elongating your neck and extending your elbows throughout the movement

PLOUGH

BEST FOR

- iliacus
- iliopsoas
- obliquus externus
- obliquus internus
- rectus abdominis
- sartorius
- tibialis anterior
- transversus abdominis

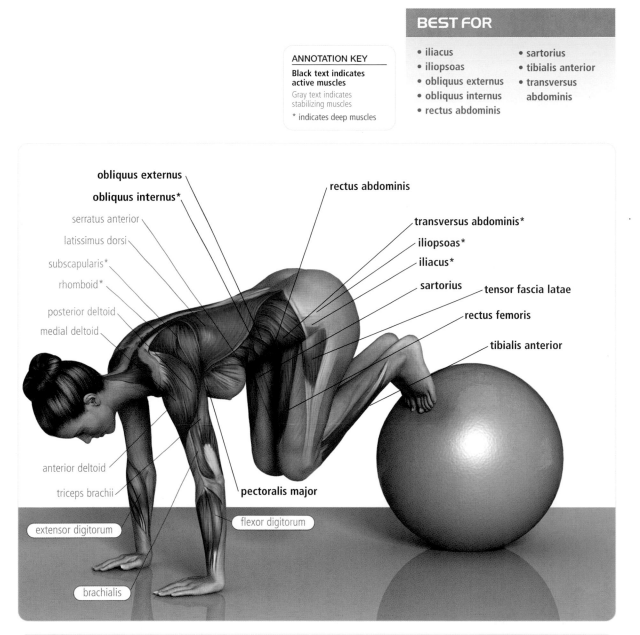

- obliquus externus
- obliquus internus*
- serratus anterior
- latissimus dorsi
- subscapularis*
- rhomboid*
- posterior deltoid
- medial deltoid
- anterior deltoid
- triceps brachii
- extensor digitorum
- brachialis
- pectoralis major
- flexor digitorum
- rectus abdominis
- transversus abdominis*
- iliopsoas*
- iliacus*
- sartorius
- tensor fascia latae
- rectus femoris
- tibialis anterior

MODIFICATIONS

More Difficult: Push your hips toward the ceiling, keeping your knees straight. The soles of your feet should be on the ball, with your head and chest between your arms and your legs and torso at a 90 degree angle. Avoid rounding your back, bending your knees, or bending your elbows.

BACK

The muscles of the back are used to move not only the spine, but also the hips, head, arms and pelvis.

There are three groups of back muscles: lower, upper and deep spinal. Lower back muscles work with hip muscles to tilt the pelvis backward and forward, as well as flex and extend the lower spine. This combination of muscles keeps the spine in a normal lordosis, without which neither the upper or lower body could function properly or efficiently. The upper back muscles depress, elevate, and rotate the scapula, and retract, rotate, adduct and abduct the humerus. These muscles are primarily responsible for all "pulling" movements, along with the biceps. The deep spinal muscles both move and stabilize the vertebrae. There are no movements in this book, or in everyday life for that matter, that do not require some contribution from the muscles of your back.

BACK ANATOMY

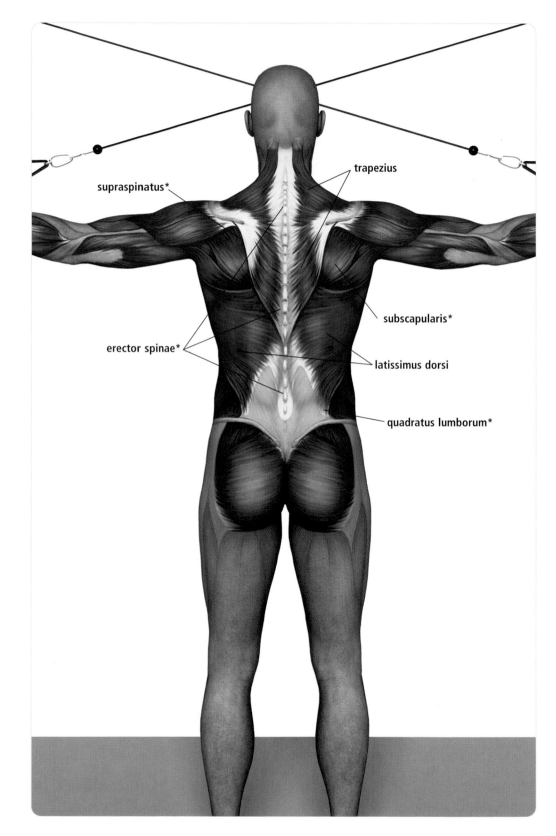

supraspinatus*

trapezius

subscapularis*

erector spinae*

latissimus dorsi

quadratus lumborum*

ANNOTATION KEY

* indicates deep muscles

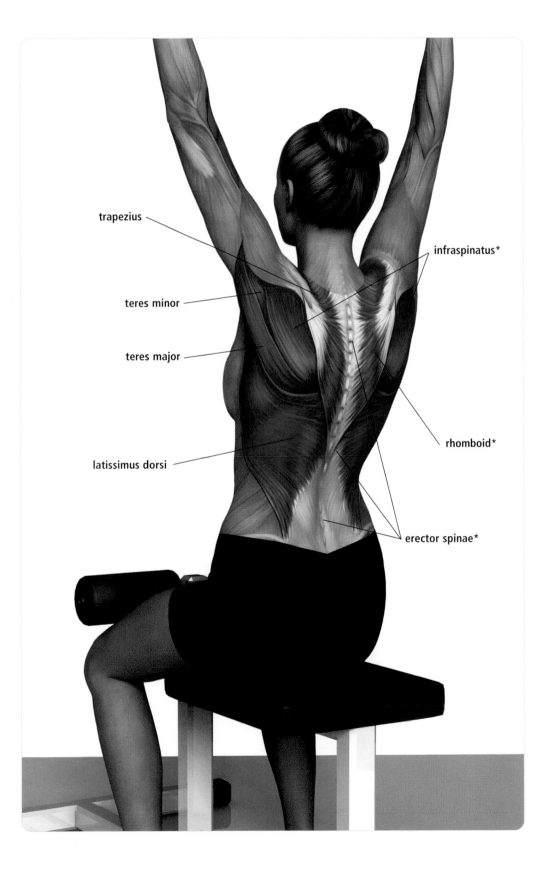

trapezius

infraspinatus*

teres minor

teres major

rhomboid*

latissimus dorsi

erector spinae*

BACK EXTENSION

Starting Position: Rest your upper legs on pads so that your hip bones are slightly above the top of the pads. Your legs and knees should be straight and contracted, with the calves on your lower legs securely pushed back into the lower pad, with your feet flat. Your torso should form a straight line from head to toe, with your spine in a neutral position. Clasp your fingers behind your head, with your shoulders down, your chin up, and your elbows pointing directly out from your shoulders.

Action: Drop your entire torso forward from your hips, keeping your spine solid and your elbows back. Contract your hamstrings, gluteals, and back to return to the starting position. Inhale as you drop forward and exhale as you return to the starting position.

LOOK FOR
- Your body to move forward from the lower gluteal, not your back or spine
- Your lower body to remain taut throughout the movement
- Your lower legs to be in continuous contact with the support pads

AVOID
- Elevating your shoulders
- Bending your knees
- Improper positioning (your hip bones should be in direct contact with the pads)

Movement Path: Your entire upper body moves from a vertical position in a downward arc and returns.

STABILIZE BY
- Contracting all of your leg muscles
- Keeping your chest high, your shoulders down, and your head neutral
- Keeping your abdomen pulled up and in

BACK EXTENSION

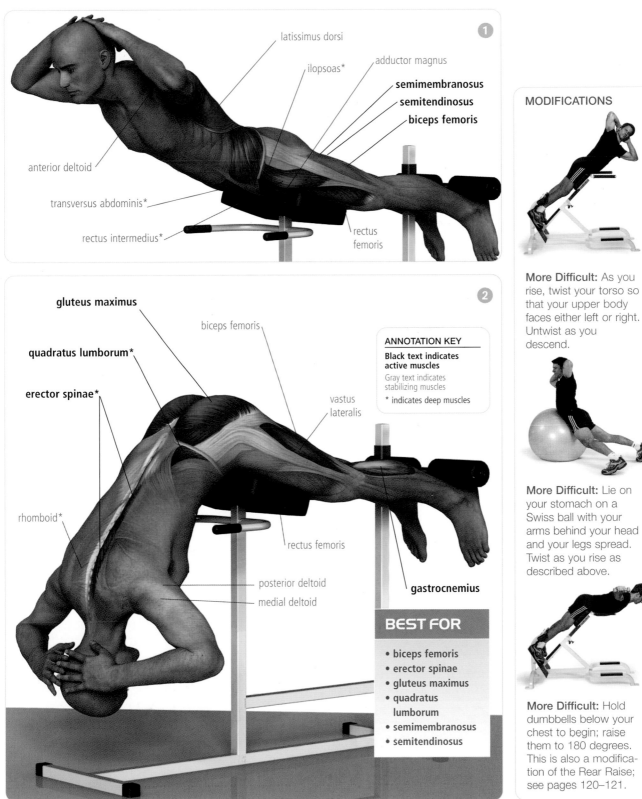

1

latissimus dorsi

ilopsoas*

adductor magnus

semimembranosus

semitendinosus

biceps femoris

anterior deltoid

transversus abdominis*

rectus intermedius*

rectus femoris

2

gluteus maximus

biceps femoris

quadratus lumborum*

erector spinae*

vastus lateralis

rhomboid*

rectus femoris

posterior deltoid

medial deltoid

gastrocnemius

ANNOTATION KEY

Black text indicates active muscles

Gray text indicates stabilizing muscles

* indicates deep muscles

BEST FOR

- biceps femoris
- erector spinae
- gluteus maximus
- quadratus lumborum
- semimembranosus
- semitendinosus

MODIFICATIONS

More Difficult: As you rise, twist your torso so that your upper body faces either left or right. Untwist as you descend.

More Difficult: Lie on your stomach on a Swiss ball with your arms behind your head and your legs spread. Twist as you rise as described above.

More Difficult: Hold dumbbells below your chest to begin; raise them to 180 degrees. This is also a modification of the Rear Raise; see pages 120–121.

ROMANIAN DEADLIFT

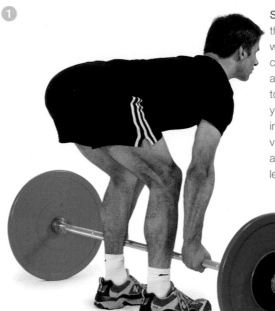

Starting Position: With the barbell on the ground and your feet shoulder-width apart, stand so that your shins contact the bar. Grasp the bar with an alternating grip (with one palm facing toward you and the other away). Bend your knees slightly, keeping your spine in a neutral position and your hips elevated, so that your head, shoulders, and hips are in a straight line and parallel to the floor.

Action: Create tension from your hands through the back of your body, all the way to your heels. Drive your back up and your hips forward, drawing the bar in a straight line vertically adjacent to your shins, and continue until you are in a full standing position.

Movement Path: Your center of mass moves vertically upward as the line of your torso rotates in an arc.

LOOK FOR
- All movement to happen at the same time
- Your spine to remain completely stable from hips to head
- Your head to be up, with your eyes forward and looking upward

AVOID
- Allowing your spine to round (by flexing forward) or change position in segments as it moves
- Bending so that your hips are above your shoulders during the movement
- Bending your elbows or shrugging your shoulders
- Allowing your weight to rest in the front part of the foot or the bar to be forward of the toe line

STABILIZE BY
- Keeping your rib cage high, your head up, and your shoulders down and back, with your shoulder blades flat on your rib cage

ROMANIAN DEADLIFT

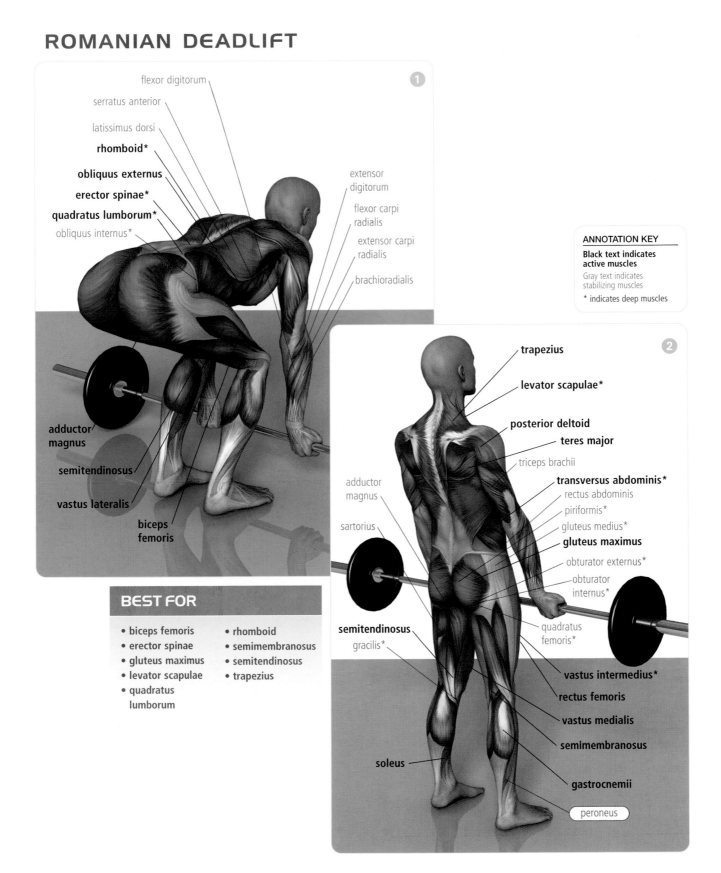

①

flexor digitorum

serratus anterior

latissimus dorsi

rhomboid*

obliquus externus

erector spinae*

quadratus lumborum*

obliquus internus*

extensor digitorum

flexor carpi radialis

extensor carpi radialis

brachioradialis

adductor magnus

semitendinosus

vastus lateralis

biceps femoris

②

trapezius

levator scapulae*

posterior deltoid

teres major

triceps brachii

transversus abdominis*

rectus abdominis

piriformis*

gluteus medius*

gluteus maximus

obturator externus*

obturator internus*

adductor magnus

sartorius

quadratus femoris*

semitendinosus

gracilis*

vastus intermedius*

rectus femoris

vastus medialis

semimembranosus

soleus

gastrocnemii

peroneus

ANNOTATION KEY

Black text indicates active muscles

Gray text indicates stabilizing muscles

* indicates deep muscles

BEST FOR

- biceps femoris
- erector spinae
- gluteus maximus
- levator scapulae
- quadratus lumborum
- rhomboid
- semimembranosus
- semitendinosus
- trapezius

ONE ARM DUMBBELL ROW

Starting Position: Place one knee and one hand on a bench, with your spine horizontal and parallel to the bench. Your knee should be directly below your hip and your hand directly below your shoulder. Place your opposite foot on the ground, bending your knee slightly. Your feet should be slightly wider than shoulder width. Grasp a dumbbell in your free hand and allow your arm to hang perpendicular to the floor.

1

LOOK FOR
- The dumbbell to move from your shoulder toward the mid-torso as it rises
- The shoulder on the working side to stay down and away from your ear

AVOID
- Any spinal movement
- Any hip rotation
- Lifting the dumbbell in a straight line
- Allowing your elbow to migrate away from the side of your torso during any part of the movement

Action: Exhale and drive your elbow directly up toward the ceiling by retracting your shoulder blade and flexing the elbow joint. Inhale and lower your elbow, keeping the rest of your body motionless.

Movement Path: Your lower arm moves from an extended position, perpendicular to the torso and pointing directly down toward the floor, straight up, remaining perpendicular to the torso in the vertical plane throughout the movement. Your upper arm moves from the beginning position upward and backward in an arc, past the torso until your hand is adjacent to your ribcage.

2

STABILIZE BY
- Keeping your hips and shoulders even
- Bearing your weight evenly on all three contact points (the weight-bearing hand, the knee on the bench, and the foot on the floor)
- Keeping your chest high and maintaining a neutral spinal position

ONE ARM DUMBBELL ROW

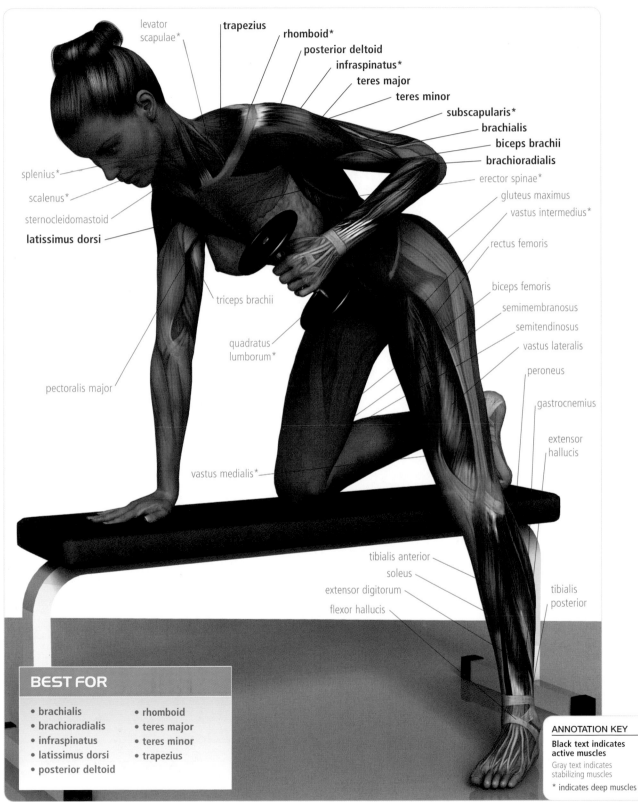

levator scapulae*
trapezius
rhomboid*
posterior deltoid
infraspinatus*
teres major
teres minor
subscapularis*
brachialis
biceps brachii
brachioradialis
splenius*
scalenus*
sternocleidomastoid
latissimus dorsi
erector spinae*
gluteus maximus
vastus intermedius*
rectus femoris
biceps femoris
semimembranosus
semitendinosus
vastus lateralis
peroneus
gastrocnemius
extensor hallucis
triceps brachii
quadratus lumborum*
pectoralis major
vastus medialis*
tibialis anterior
soleus
extensor digitorum
flexor hallucis
tibialis posterior

BEST FOR

- brachialis
- brachioradialis
- infraspinatus
- latissimus dorsi
- posterior deltoid
- rhomboid
- teres major
- teres minor
- trapezius

ANNOTATION KEY

Black text indicates active muscles
Gray text indicates stabilizing muscles
* indicates deep muscles

BARBELL ROW

Starting Position: Bend over so that your back is almost parallel to the floor, with your feet hip-width apart. Hold a barbell with straight arms and your palms facing your body. Make sure your spine is in a neutral position. Your hips should be behind your feet and your knees should be bent, with your torso angled forward so that your shoulders are slightly above your hips. The barbell should be above your shoelaces. Your chest should be up, your abdomen pulled in, and your chin slightly raised.

1

STABILIZE BY
- Pulling your abdomen up and in
- Contracting your spinal muscles and maintaining your spinal position
- Keeping your knees bent at a fixed angle
- Keeping your weight toward the heel of your feet and the barbell close to your legs throughout the movement
- Contracting your gluteals and hamstrings throughout the movement
- Keeping your chest elevated

LOOK FOR
- The bar to travel in a straight vertical path
- Your hips and torso to remain motionless, with your head up

AVOID
- Any spinal movement
- Rounding your shoulders forward
- Allowing the bar to descend in a path that is in front of toe line
- Changing either your knee or hip position during the movement
- Holding your breath

Action: Exhale and pull the barbell directly upward to your mid-torso, just below your ribcage. Your forearms travel vertically and your elbows are pulled directly upward, above the plane of your back. Squeeze your shoulders down and back. Inhale and lower the bar in the same path, close to and parallel with your lower legs.

Movement Path: Your center of mass remains stationary. Your upper arms move from a forward, downward-facing position backward and upward, while your lower arms remain forward and downward-facing.

2

BARBELL ROW

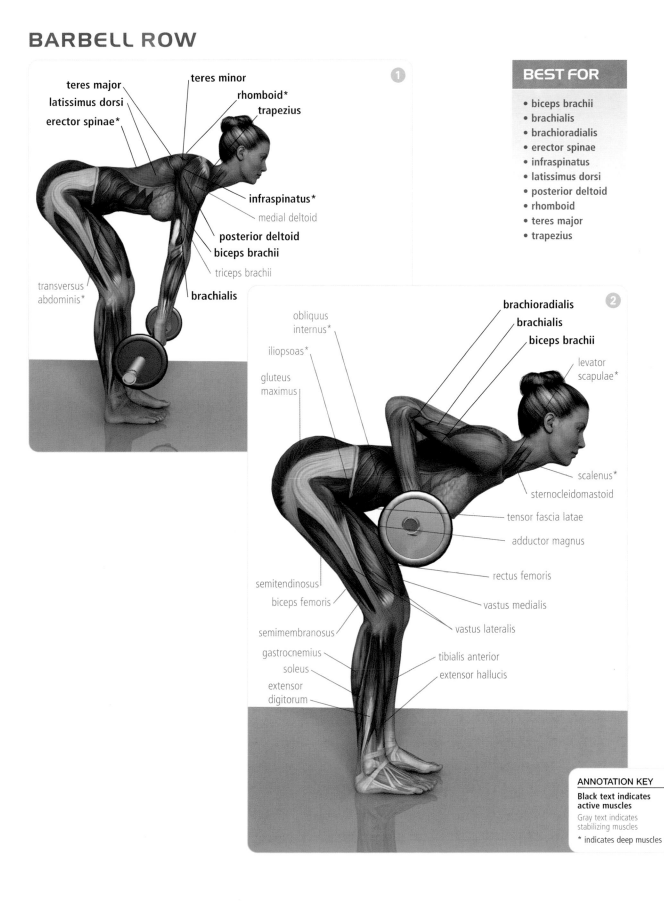

1

teres major
latissimus dorsi
erector spinae*
teres minor
rhomboid*
trapezius
infraspinatus*
medial deltoid
posterior deltoid
biceps brachii
triceps brachii
transversus abdominis*
brachialis

2

obliquus internus*
iliopsoas*
gluteus maximus
brachioradialis
brachialis
biceps brachii
levator scapulae*
scalenus*
sternocleidomastoid
tensor fascia latae
adductor magnus
rectus femoris
vastus medialis
vastus lateralis
semitendinosus
biceps femoris
semimembranosus
gastrocnemius
soleus
extensor digitorum
tibialis anterior
extensor hallucis

BEST FOR

- biceps brachii
- brachialis
- brachioradialis
- erector spinae
- infraspinatus
- latissimus dorsi
- posterior deltoid
- rhomboid
- teres major
- trapezius

ANNOTATION KEY

Black text indicates active muscles
Gray text indicates stabilizing muscles
* indicates deep muscles

CHIN-UP

Starting Position: Hanging from a bar, with your knees bent only very slightly, grip the bar with your palms in (i.e., facing your body). Keep your head in a neutral alignment. You hands should be shoulder-width apart.

Action: Pull your body up vertically until your upper chest is aligned with the bar: this is the end of the concentric phase. Lower your body back down to the starting position with your elbows fully extended (the end of the eccentric phase).

LOOK FOR
- Your arms to return to a full extension
- You shoulder blades to draw together and downward at the beginning of the movement

AVOID
- Swinging, jerking, chin "pecking," or hyperextension of elbows

Movement Path: Your body moves vertically up. Your upper body tilts back slightly to allow your chin to smoothly pass the bar line.

STABILIZE BY
- Retracting your scapula
- Keeping your core tight to prevent swinging

CHIN-UP

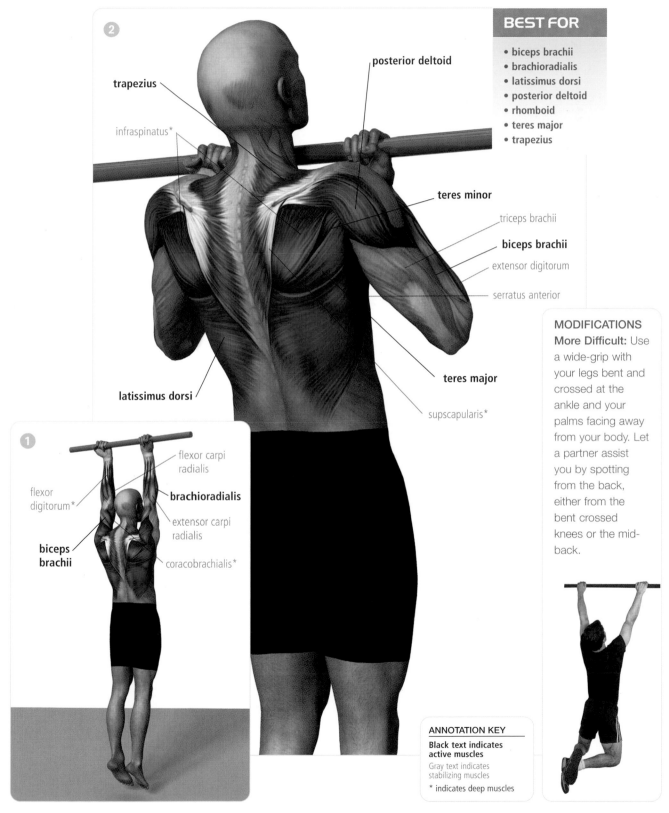

2

trapezius

infraspinatus*

posterior deltoid

teres minor

triceps brachii

biceps brachii

extensor digitorum

serratus anterior

teres major

supscapularis*

latissimus dorsi

1

flexor carpi radialis

flexor digitorum*

brachioradialis

extensor carpi radialis

biceps brachii

coracobrachialis*

BEST FOR

- biceps brachii
- brachioradialis
- latissimus dorsi
- posterior deltoid
- rhomboid
- teres major
- trapezius

MODIFICATIONS

More Difficult: Use a wide-grip with your legs bent and crossed at the ankle and your palms facing away from your body. Let a partner assist you by spotting from the back, either from the bent crossed knees or the mid-back.

ANNOTATION KEY

Black text indicates active muscles

Gray text indicates stabilizing muscles

* indicates deep muscles

BODY ROW

Starting Position: Hang from a bar with your body in a flat plane. The line of your body should be at a 45 degree angle to the floor. Grasp the bar with both arms in supine or prone grips. Your elbows should be at 90 degree angles.

STABILIZE BY
- Fixing your shoulders in one position
- Locking your knees
- Keeping your ankles in a fixed position
- Keeping your hips, abdominal muscles, and lower back rigid

LOOK FOR
- A single plane of movement, maintaining a straight line from your head to your ankles

AVOID
- A segmental elevation, such as your shoulders rising before hips or vice versa
- Elevating your shoulders toward your ears
- Moving your head forward

Action: Move your feet away from the bar until your arms are straight, keeping on your heels. Pull your body toward the bar until your chest touches it. Lower yourself slowly and repeat. The bottom of your chest should always touch the bar at the end of the movement. Keep your body in a straight line on your heels.

Movement Path: Your entire body moves in a single arc with your feet as the fulcrum.

BODY ROW

BEST FOR

- biceps brachii
- brachialis
- brachioradialis
- infraspinatus
- latissimus dorsi
- rhomboid
- teres major
- teres minor
- trapezius

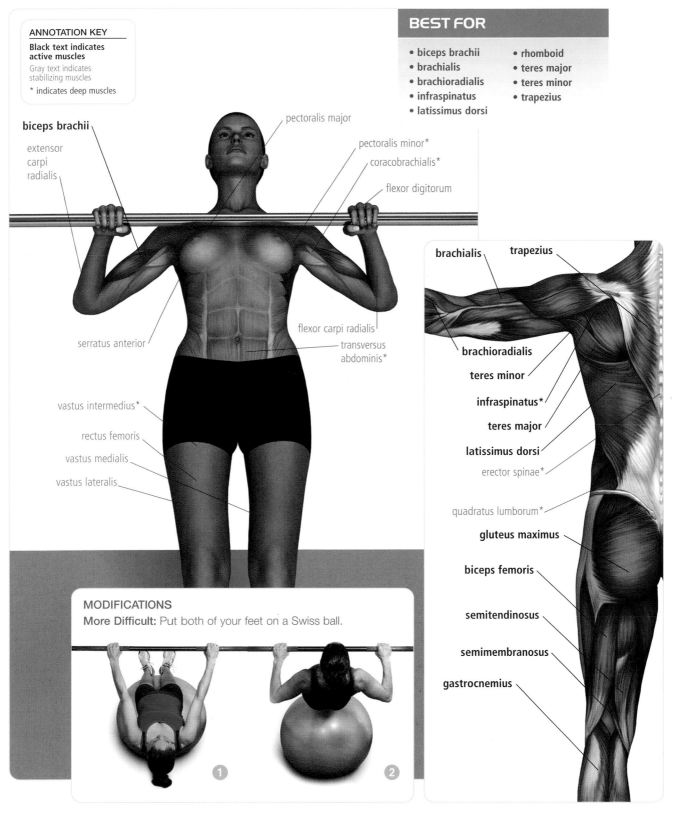

biceps brachii

extensor carpi radialis

serratus anterior

vastus intermedius*

rectus femoris

vastus medialis

vastus lateralis

pectoralis major

pectoralis minor*

coracobrachialis*

flexor digitorum

flexor carpi radialis

transversus abdominis*

brachialis

trapezius

brachioradialis

teres minor

infraspinatus*

teres major

latissimus dorsi

erector spinae*

quadratus lumborum*

gluteus maximus

biceps femoris

semitendinosus

semimembranosus

gastrocnemius

MODIFICATIONS

More Difficult: Put both of your feet on a Swiss ball.

CABLE ROW

Starting Position: Sit on the bench with your legs extended in front of you. Keep your legs slightly bent and reach forward to grab the handle with both of your hands in a neutral grip (so that your palms face each other). Start with your hands over your ankles, your back arched, and your knees slightly bent.

STABILIZE BY
- Using your abdominal and back muscles to maintain an upright posture

LOOK FOR
- A slow continuous movement in the horizontal plane

AVOID
- Rounding your back
- Too much movement in your torso (i.e., leaning forward or backward too much)
- Elevating your shoulders

Action: As you pull the handle toward your midsection, squeezing your shoulder blades back and down, lean backward (keeping your back arched) to about 5 degrees past perpendicular. To repeat, lean forward to about 5 degrees less than perpendicular and slowly straighten your arms, then pull the handle toward your torso again as you lean back to 5 degrees past perpendicular. The set should be done within this 10 degree range. Make sure you keep your legs slightly bent, your back arched, and your elbows shoulder-width apart.

Movement Path: The movement of the bar is horizontal; your torso oscillates in a narrow 10 degree arc.

CABLE ROW

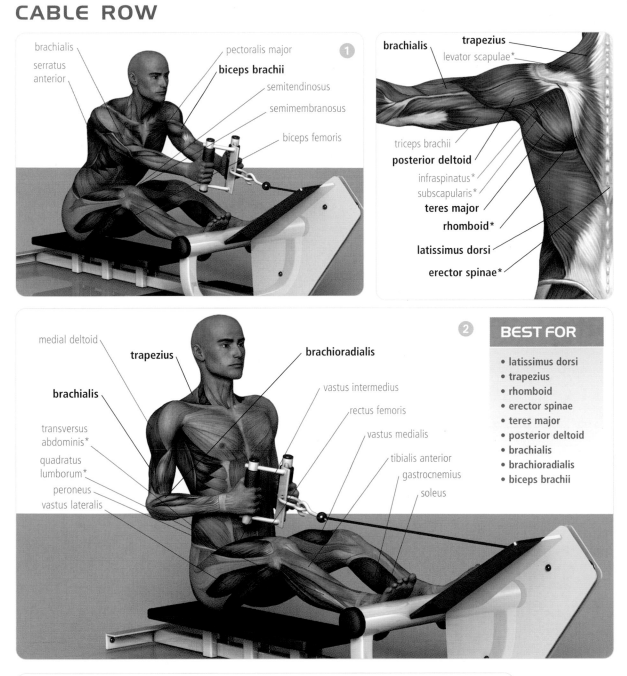

①

brachialis
serratus anterior
pectoralis major
biceps brachii
semitendinosus
semimembranosus
biceps femoris

brachialis
trapezius
levator scapulae*
triceps brachii
posterior deltoid
infraspinatus*
subscapularis*
teres major
rhomboid*
latissimus dorsi
erector spinae*

②

medial deltoid
trapezius
brachioradialis
brachialis
vastus intermedius
rectus femoris
vastus medialis
transversus abdominis*
quadratus lumborum*
tibialis anterior
gastrocnemius
peroneus
soleus
vastus lateralis

BEST FOR

- latissimus dorsi
- trapezius
- rhomboid
- erector spinae
- teres major
- posterior deltoid
- brachialis
- brachioradialis
- biceps brachii

MODIFICATIONS
Similar Difficulty:
Change your grip position to either supinated (with your palms up) or pronated (with your palms down).

① **②**

ANNOTATION KEY

Black text indicates active muscles
Gray text indicates stabilizing muscles

* indicates deep muscles

STRAIGHT ARM PULL-DOWN

Starting Position: Stand, facing the high pulley, with your legs shoulder-width apart and your spine in a neutral position. Grasp the bar in an overhand grip with your palms facing down and your arms extended.

LOOK FOR
- A neutral spine
- Retrated scapula

AVOID
- Elevating your shoulders toward your ears
- Arching your back
- Bending your arms

Action: Pull the bar straight down toward your lap, bringing your shoulders down and back, so that your plams face your thighs at the bottom of the movement. Return the attachment and repeat.

Movement Path: Your arms should extend slightly beyond your body in one plane of movement. Your torso rotates around your center of mass at the same rate as you pull the cable.

STABILIZE BY
- Drawing in your navel
- Retracting your shoulder blades
- Keeping your shoulders in one position and your upper arms alongside your body

STRAIGHT ARM PULL-DOWN

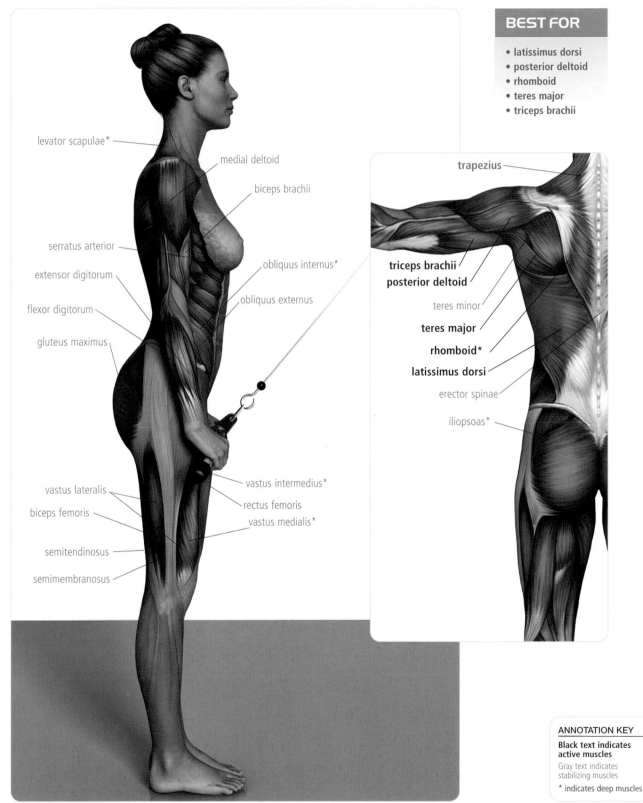

BEST FOR

- latissimus dorsi
- posterior deltoid
- rhomboid
- teres major
- triceps brachii

levator scapulae*

medial deltoid

biceps brachii

serratus arterior

obliquus internus*

extensor digitorum

obliquus externus

flexor digitorum

gluteus maximus

vastus intermedius*

rectus femoris

vastus medialis*

vastus lateralis

biceps femoris

semitendinosus

semimembranosus

trapezius

triceps brachii
posterior deltoid

teres minor

teres major

rhomboid*

latissimus dorsi

erector spinae

iliopsoas*

ANNOTATION KEY

**Black text indicates
active muscles**

Gray text indicates
stabilizing muscles

* indicates deep muscles

CABLE REAR RAISE

Starting Position:
Standing equidistant and directly between two high cables, grasp the handles with the opposing hands so that your palms face each other. Take two full steps backward and extend your arms so that your hands land directly in front of your shoulder joints with your arms completely straight, parallel to each other and the ground. The cable should be crossed between your hands at approximately chin height.

1

LOOK FOR
- Your arms to be fully extended throughout the movement

AVOID
- Bending your elbows
- Shrugging your shoulders
- Changing the plane of your arm movement

STABILIZE BY
- Keeping your chest high and your abdomen pulled up and in
- Keeping your shoulders down and back throughout the movement
- Keeping your shoulders, hips, knees, and feet in alignment

Action: Exhale and retract your shoulders and arms backward and out to the side until your arms are pointing directly out. Make sure your arms remain in a horizontal plane. Inhale and allow your arms to return in the same but opposite movement.

Movement Path: Your torso, hips, and legs remain stationary while your arms move in a horizontal, 90 degree arc from the front of your body out to both sides.

2

CABLE REAR RAISE

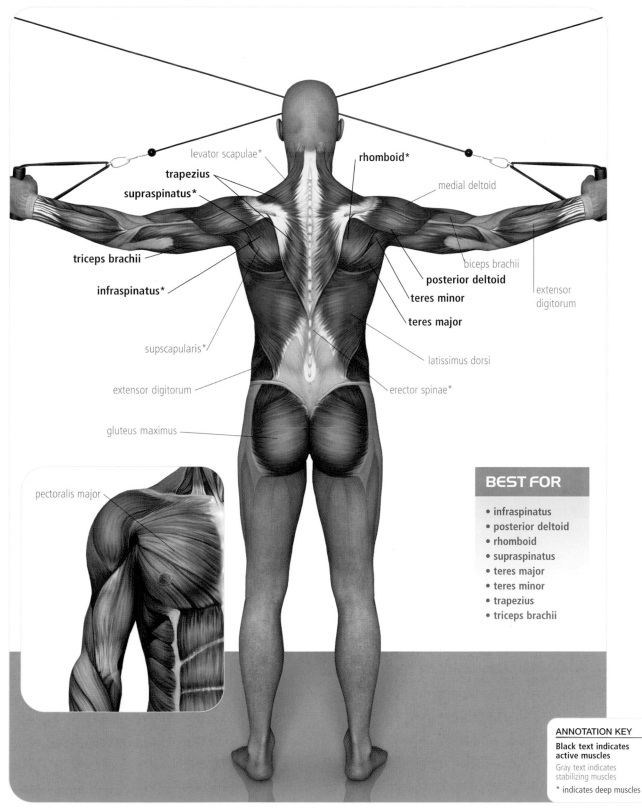

levator scapulae*

rhomboid*

medial deltoid

trapezius

supraspinatus*

triceps brachii

infraspinatus*

biceps brachii

posterior deltoid

teres minor

teres major

extensor digitorum

supscapularis*

latissimus dorsi

extensor digitorum

erector spinae*

gluteus maximus

pectoralis major

BEST FOR

- infraspinatus
- posterior deltoid
- rhomboid
- supraspinatus
- teres major
- teres minor
- trapezius
- triceps brachii

ANNOTATION KEY

Black text indicates active muscles
Gray text indicates stabilizing muscles
* indicates deep muscles

CHEST

The muscles of the chest originate on the collar bone and sternum and insert on the upper arm. They are responsible for adduction, internal rotation and forward flexion of the humerus. This muscle group is responsible for "pushing" movements and interacts synergistically with the anterior deltoid of the shoulder and triceps of the arm to accomplish this.

CHEST ANATOMY

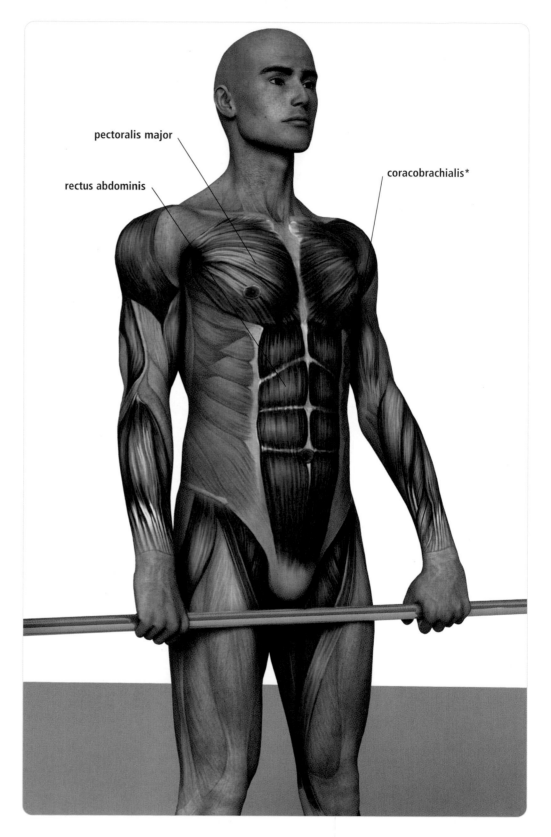

pectoralis major

rectus abdominis

coracobrachialis*

ANNOTATION KEY

* indicates deep muscles

coracobrachialis*

pectoralis major

pectoralis minor*

serratus anterior

INCLINE DUMBBELL FLY

Starting Position: Sit on an inclined bench with your shoulders higher than your hips at no greater than a 60 degree angle. Hold the dumbbells above your chest so that your palms face each other. Your elbows should be very slightly bent. Your shoulder blades should contact the bench. Keep your spine in a neutral position and your feet flat on the floor.

Action: Keeping your elbows bent, push your hands apart and inhale until your hands drop to just below the height of your chest. Return to the starting position by squeezing your chest and bringing the weights back to the starting position along the same path as the descent exhaling as you do so.

LOOK FOR
- Your chest and rib cage to rise as the weight descends
- Your spine and shoulders to remain in the same position as you return to the starting position

AVOID
- Moving your head or chin forward off of the bench
- Elevating your shoulders toward your ears
- Bending your elbows excessively as the weight descends or flattening them as the weight ascends

Movement Path: As the dumbbells leave the starting position they come down and apart in an arc that begins vertically and drops to horizontal relative to the ground, not the bench.

STABILIZE BY
- Keeping your grip strong and your upper arms (both biceps and triceps) contracted.
- Ensuring that your shoulder blades remain in contact with the bench throughout the movement.
- Keeping your feet flat and your neck long.

INCLINE DUMBBELL FLY

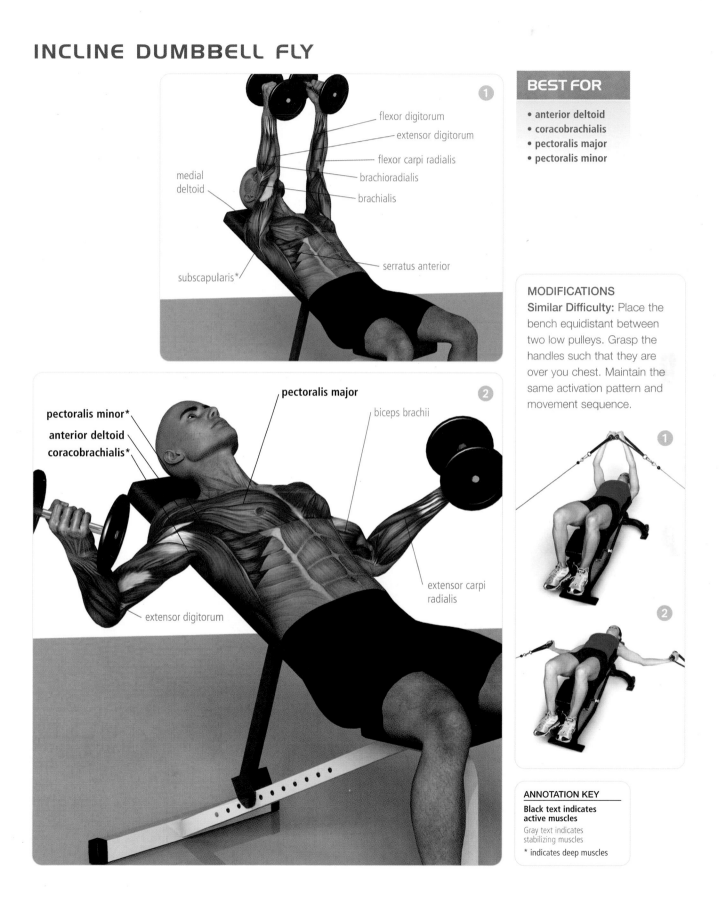

1

flexor digitorum

extensor digitorum

flexor carpi radialis

brachioradialis

medial deltoid

brachialis

serratus anterior

subscapularis*

2

pectoralis minor*

pectoralis major

anterior deltoid

biceps brachii

coracobrachialis*

extensor carpi radialis

extensor digitorum

BEST FOR

- anterior deltoid
- coracobrachialis
- pectoralis major
- pectoralis minor

MODIFICATIONS

Similar Difficulty: Place the bench equidistant between two low pulleys. Grasp the handles such that they are over you chest. Maintain the same activation pattern and movement sequence.

1

2

ANNOTATION KEY

Black text indicates active muscles

Gray text indicates stabilizing muscles

* indicates deep muscles

BENCH PRESS

Starting Position: Lie on a bench, holding a barbell above your chest so that your arms are directly above your collarbone. Your arms should be extended, your shoulder blades and feet flat, and your spine in a neutral position. Grip the bar so that your hands are wider than your shoulder width in an overhand grip, with your thumb grasping the bar underneath.

Action: Inhale and lower the barbell to your chest, finishing on or slightly above the nipple line. Exhale, extending your arms to push up to the ceiling until returned to the starting position.

1

LOOK FOR
- Your rib cage to remain open and rise during the descent phase
- Your shoulders to remain retracted and away from your ears during the ascent

AVOID
- Dropping the weight quickly
- Bouncing the bar off your chest
- Changing your spinal position (either arching or flattening your back) during the movement
- Picking your feet up off the floor

Movement Path: As the bar descends, it remains horizontal. There is a slight movement downward on the torso, from the clavicle toward the mid-lower chest, which returns on the upward movement.

STABILIZE BY
- Keeping your spine neutral and your forearms perpendicular to the bar (directly below it) throughout the movement
- Ensuring that your shoulders, head, and hips remain in contact with the bench at all times

2

BENCH PRESS

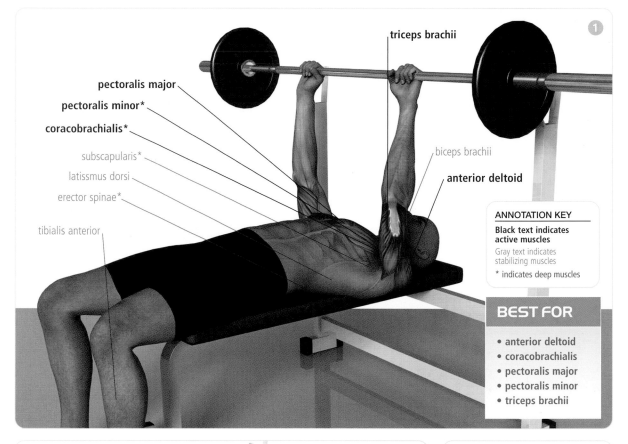

triceps brachii

pectoralis major

pectoralis minor*

coracobrachialis*

subscapularis*

latissmus dorsi

erector spinae*

tibialis anterior

biceps brachii

anterior deltoid

ANNOTATION KEY
Black text indicates active muscles
Gray text indicates stabilizing muscles
* indicates deep muscles

BEST FOR

- anterior deltoid
- coracobrachialis
- pectoralis major
- pectoralis minor
- triceps brachii

trapezius

iliopsoas*

soleus

MODIFICATIONS
More Difficult: Grip the bar with your hands closer together. Maintain the same activation pattern and movement sequence.

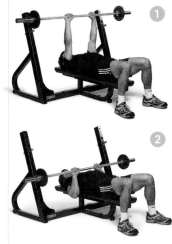

INCLINE BARBELL PRESS

1 **Starting Position:** Lie on a bench at a 45 degree angle, holding a barbell above your chest so that in profile, your hands, arms, and shoulders are vertical. Your arms should be extended, your shoulder blades and feet flat, and your spine in a neutral position. Grip the bar so that your hands are wider than your shoulder width in an overhand grip, with your thumb grasping the bar underneath.

Action: Inhale and lower the barbell until it touches your upper chest, just below your collarbone. Exhale, extending your arms to push up to the ceiling until returned to the starting position.

LOOK FOR
- Your rib cage to remain open and rise during the descent phase
- Your shoulders to remain retracted and away from your ears during the ascent

AVOID
- Dropping the weight quickly
- Bouncing the bar off your chest
- Changing your spinal position (either arching or flattening your back) during the movement
- Picking your feet up off the floor

2 **Movement Path:** As the bar descends, it remains horizontal. Your elbows and forearms move directly downward in a vertical plane.

STABILIZE BY
- Keeping your spine neutral and your forearms perpendicular to the bar (directly below it) throughout the movement
- Ensuring that your shoulders, head, and hips remain in contact with the bench at all times
- Keeping your feet in contact with the floor

INCLINE BARBELL PRESS

trapezius

teres major

subscapularis*

rhomboid*

triceps brachii

pectoralis
minor*

brachioradialis

anterior deltoid

pectoralis major

latissimus
dorsi

serratus anterior

flexor digitorum

extensor carpi radialis
flexor carpi radialis
extensor carpi radialis

coracobrachialis*

BEST FOR

- anterior deltoid
- coracobrachialis
- pectoralis major
- pectoralis minor
- triceps brachii

ANNOTATION KEY

**Black text indicates
active muscles**

Gray text indicates
stabilizing muscles

* indicates deep muscles

DIP

Starting Position: With your feet on the assist bar, grasp the handles and extend your arms completely, with your hands adjacent to your hips and your torso leaning slightly forward. Make sure that your shoulders are down and away from your ears and that spine is in a neutral position.

LOOK FOR
• Your torso to lean forward as you descend

AVOID
• Looking down
• Shifting your elbows to a position that is not directly over your hands

Action: Lower your body until your upper arms are parallel to the floor. Your knees should be slightly behind your hips and your chest slightly in front of them. Return to the starting position by extending your elbows.

Movement Path: As your torso drops, there is a slight slide backward from your knees; your center of mass drops directly downward.

STABILIZE BY
• Arching your lower back slightly and keeping it solid throughout the movement
• Placing your elbows directly over your hands
• Keeping your chin up and your eyes looking directly forward

DIP

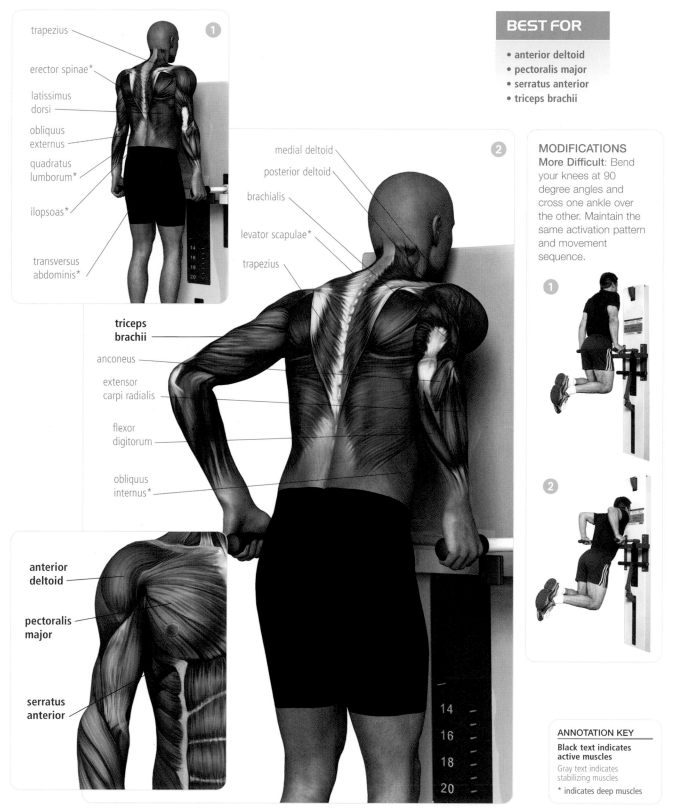

trapezius

erector spinae*

latissimus dorsi

obliquus externus

quadratus lumborum*

ilopsoas*

transversus abdominis*

medial deltoid

posterior deltoid

brachialis

levator scapulae*

trapezius

triceps brachii

anconeus

extensor carpi radialis

flexor digitorum

obliquus internus*

anterior deltoid

pectoralis major

serratus anterior

MODIFICATIONS
More Difficult: Bend your knees at 90 degree angles and cross one ankle over the other. Maintain the same activation pattern and movement sequence.

ANNOTATION KEY

Black text indicates active muscles

Gray text indicates stabilizing muscles

* indicates deep muscles

MACHINE CHEST PRESS

Starting Position: Sit vertically with your feet flat, either on a footstool or the floor. Keeping your spine in a neutral position, grasp the handles so that your forearms are parallel to the floor and your elbows are directly behind the handles in the same plane. Keep your shoulders down, your chin back, and your head up.

Action: Inhale and slowly allow the handles to come toward you by bending your elbows. Once your hands are parallel to your chest, pause, exhale, push the handles, and return to the starting position by extending your elbows.

Movement Path: The handles of the machine draw back in a horizontal plane toward your torso until your hands are parallel to your chest, and return to the starting position via the same path.

LOOK FOR
- Your chest to remain high, your forearms to remain parallel to the ground, and your elbows to remain directly behind your hands

AVOID
- Flattening your lower back, overextending your shoulders forward, or allowing your elbows to either elevate or drop during any portion of the exercise
- Picking your feet up

STABILIZE BY
- Keeping your tail bone, upper back, and head in contact with the bench (your lower back should be slightly off of it) throughout the movement, with your shoulders down and away from your ears

MACHINE CHEST PRESS

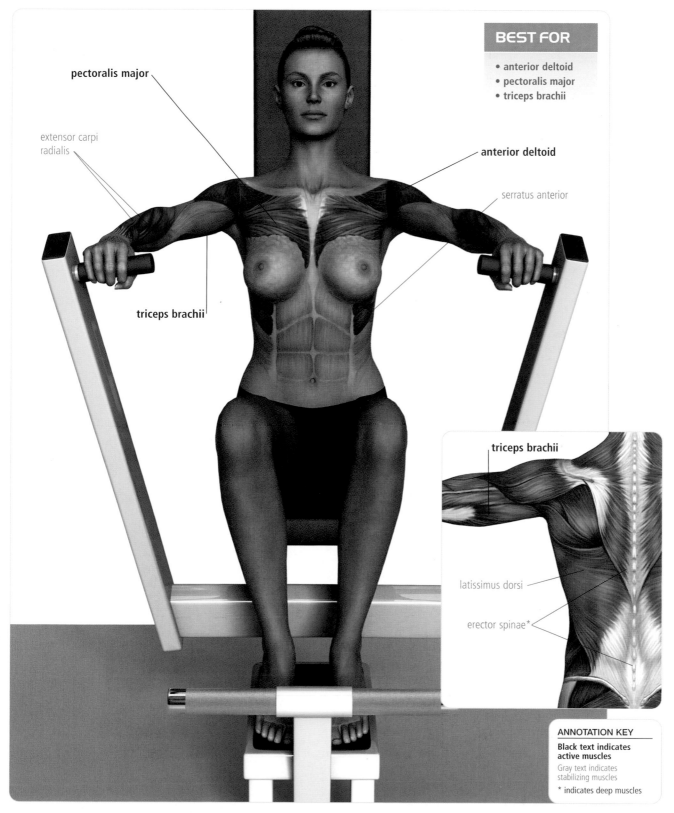

pectoralis major

extensor carpi
radialis

triceps brachii

BEST FOR

- anterior deltoid
- pectoralis major
- triceps brachii

anterior deltoid

serratus anterior

triceps brachii

latissimus dorsi

erector spinae*

ANNOTATION KEY

**Black text indicates
active muscles**

Gray text indicates
stabilizing muscles

* indicates deep muscles

SHOULDERS

Because the structure of the shoulder joint is inherently mobile and less stable than the hip or knee joints, there are many muscles that traverse it from multiple directions in order to maintain stability. They are responsible for the depression and elevation of the scapula, and take part in moving the upper arm in every direction and stabilizing it.

In any exercise or activity that involves movement of the arms, the muscles of the shoulder play a part. Since these muscles interact directly with the torso and arm, the muscles of the chest, back, core and arms are integral to their development and function.

SHOULDERS ANATOMY

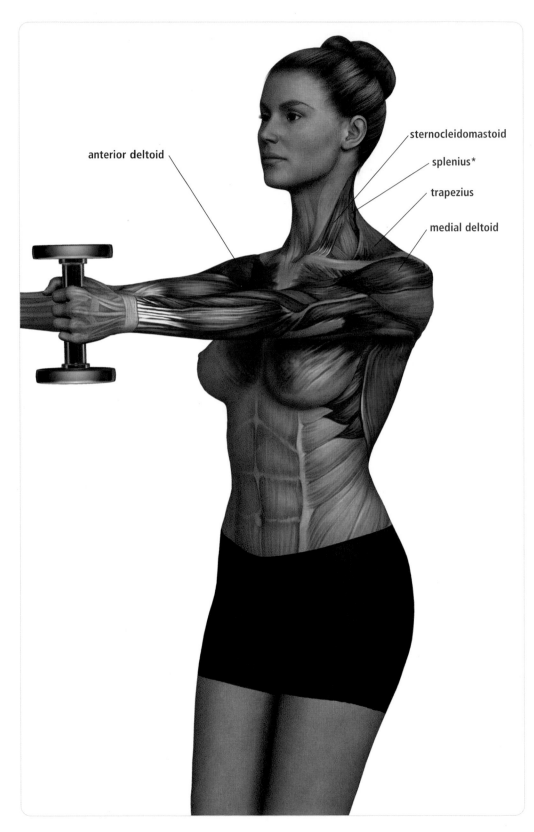

anterior deltoid

sternocleidomastoid

splenius*

trapezius

medial deltoid

ANNOTATION KEY

* indicates deep muscles

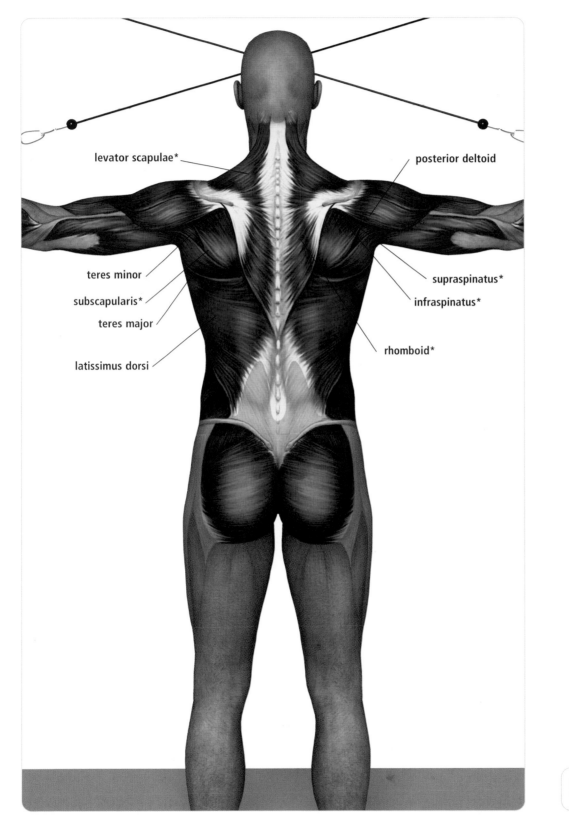

levator scapulae*

posterior deltoid

teres minor

subscapularis*

teres major

latissimus dorsi

supraspinatus*

infraspinatus*

rhomboid*

ANNOTATION KEY

* indicates deep muscles

SHOULDER EXTERNAL ROTATION

Starting Position: Lie on your side, supporting your head with your hand and keeping your bottom arm bent. Bend your knees slightly, with your spine in a neutral position and your upper arm in an exact line with your spine. Bend your elbow at 90 degrees, with the palm of the hand holding the dumbbell facing your bellybutton.

LOOK FOR
- A rotational movement around your torso, in a line parallel to your waistline

AVOID
- Raising your upper arm or elbow away from your torso
- Letting your shoulder joint slide either forward toward your chest or backward by pinching your shoulder blades

Action: Holding your elbow by your side, rotate the back of your hand and forearm toward the ceiling, until just below perpendicular or as high as possible without changing your shoulder or upper arm position.

Movement Path: Your forearm should remain at a 90 degree angle to your elbow and upper arm. Your hand moves directly upward.

STABILIZE BY
- Keeping your shoulder in one position and your upper arm held tightly to your torso

SHOULDER EXTERNAL ROTATION

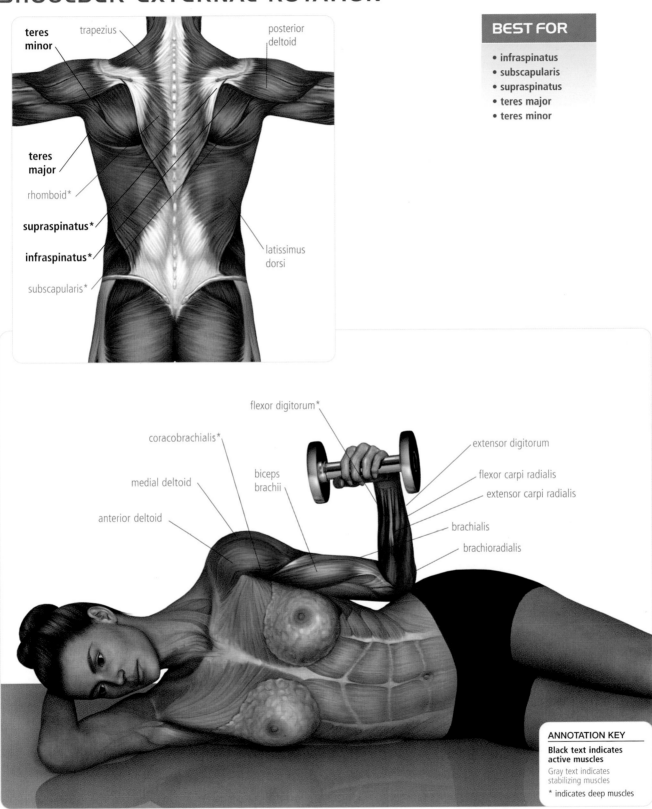

teres minor

trapezius

posterior deltoid

teres major

rhomboid*

supraspinatus*

infraspinatus*

subscapularis*

latissimus dorsi

flexor digitorum*

coracobrachialis*

medial deltoid

anterior deltoid

biceps brachii

extensor digitorum

flexor carpi radialis

extensor carpi radialis

brachialis

brachioradialis

ANNOTATION KEY

Black text indicates active muscles

Gray text indicates stabilizing muscles

* indicates deep muscles

REAR RAISE

Starting Position: Bend your knees slightly and drop your torso forward, so that your spine is just above 90 degrees from your hips, almost parallel to the floor. Keep your spine in a neutral position and your chin slightly elevated. Your shoulders should be down. Grasp the dumbbells with your hands directly under your chest toward the floor, bending your elbows slightly. **1**

STABILIZE BY
- Keeping your torso stationary
- Keeping your shoulders down and your neck long, with your head up
- Pulling your abdomen up and in
- Contracting your gluteals and hamstrings, with your spinal muscles active

LOOK FOR
- Your spine to remain in one position
- Your hands to move not only back but away (or out) from your torso in a slow and controlled fashion
- A small pause at the top of the movement

AVOID
- Any spinal, hip, knee, or head movement
- Elevating your shoulders toward your ears or extending your arms forward, toward the ground, in the bottom position
- Moving your shoulder blades from their down and flat position as your arms descend toward the floor

Action: Your arms should remain at a 90 degree angle to your torso, your knees should remain bent, and your elbows should remain in their extended position. Bring your hands to just above a horizontal line, contracting and depressing your shoulder blades. **2**

Movement Path: The dumbbells should trace a semicircle in a vertical plane.

REAR RAISE

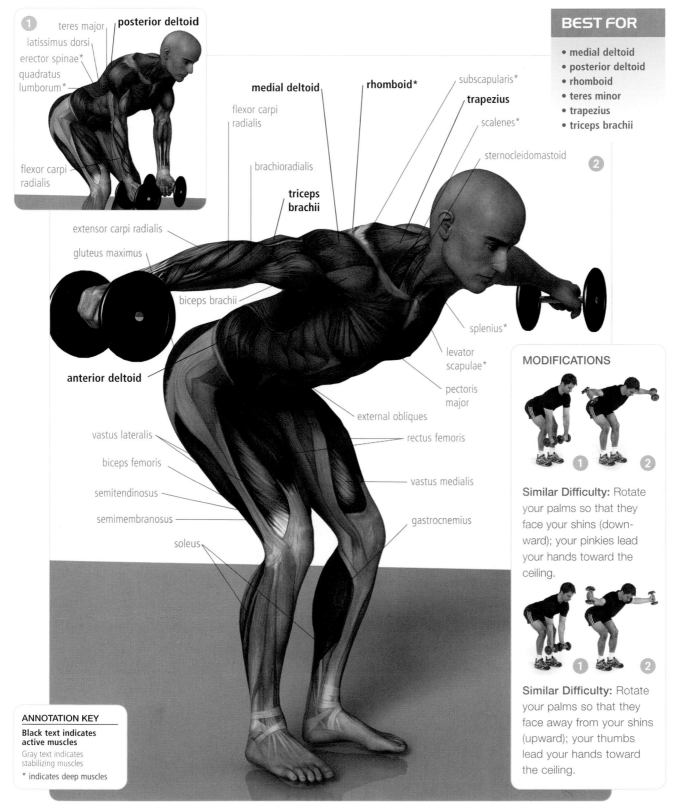

1

teres major
latissimus dorsi
erector spinae*
quadratus
lumborum*
posterior deltoid

flexor carpi
radialis

medial deltoid

rhomboid*

subscapularis*

trapezius

scalenes*

sternocleidomastoid

2

flexor carpi
radialis

brachioradialis

**triceps
brachii**

extensor carpi radialis

gluteus maximus

biceps brachii

splenius*

levator
scapulae*

anterior deltoid

pectoris
major

external obliques

vastus lateralis

rectus femoris

biceps femoris

vastus medialis

semitendinosus

semimembranosus

gastrocnemius

soleus

BEST FOR

- **medial deltoid**
- **posterior deltoid**
- **rhomboid**
- **teres minor**
- **trapezius**
- **triceps brachii**

MODIFICATIONS

1 **2**

Similar Difficulty: Rotate
your palms so that they
face your shins (down-
ward); your pinkies lead
your hands toward the
ceiling.

1 **2**

Similar Difficulty: Rotate
your palms so that they
face away from your shins
(upward); your thumbs
lead your hands toward
the ceiling.

ANNOTATION KEY

**Black text indicates
active muscles**

Gray text indicates
stabilizing muscles

* indicates deep muscles

UPRIGHT ROWS

Starting Position: Stand with your feet shoulder-width apart, grasping the dumbbells so that your palms face your body and your knuckles point away. Keep your knees slightly bent and your chest up.

LOOK FOR
- Your arms to rise simultaneously; your hands should never be higher than your elbows
- An upright posture to be maintained throughout the movement

AVOID
- An excessive shrug or rounding your shoulders
- Keeping your elbows close to your body or forward of the lateral plane of your body (from the shoulder joint)

Action: Pull your hands up until your elbows and forearms are virtually parallel to the ground. The dumbbells should end in front of your collarbones.

Movement Path: Pull your arms directly up, parallel to the line of your torso, with a slight retraction of your elbows and shoulders at the top of the movement.

STABILIZE BY
- Keeping your chest and ribcage high
- Keeping your spine solid and in a neutral position
- Bending your hips and knees slightly
- Keeping your eyes and head forward

UPRIGHT ROWS

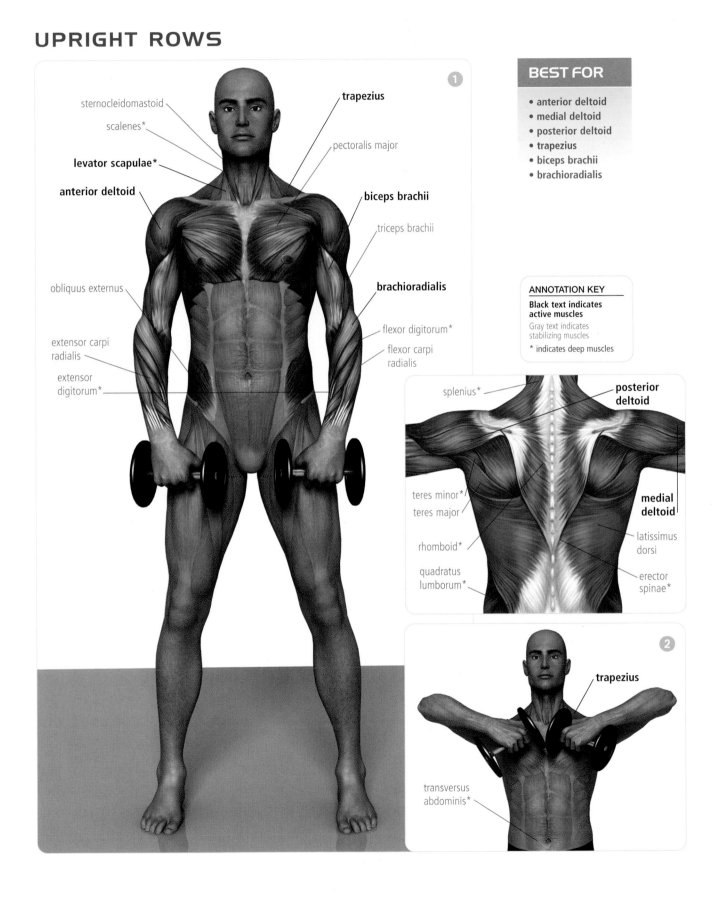

BEST FOR

- anterior deltoid
- medial deltoid
- posterior deltoid
- trapezius
- biceps brachii
- brachioradialis

ANNOTATION KEY

Black text indicates active muscles
Gray text indicates stabilizing muscles

* indicates deep muscles

sternocleidomastoid
scalenes*
levator scapulae*
anterior deltoid
obliquus externus
extensor carpi radialis
extensor digitorum*

trapezius
pectoralis major
biceps brachii
triceps brachii
brachioradialis
flexor digitorum*
flexor carpi radialis

splenius*
teres minor*
teres major
rhomboid*
quadratus lumborum*

posterior deltoid
medial deltoid
latissimus dorsi
erector spinae*

trapezius
transversus abdominis*

DUMBBELL SHRUG

Starting Position: Stand straight up, gripping dumbbells at your side.

Action: Elevate your shoulders toward your ears, keeping your head in a neutral position and exhaling as you contract and elevate.

1

2

LOOK FOR
• Movement up and in from your shoulder joints

AVOID
• Tilting your head in any direction, bending your elbows, or jutting your chin forward

Movement Path: Your shoulders move up and in toward your ears. Your elbows remain straight and your head remains in one position.

STABILIZE BY
• Keeping your head up and your chest high
• Keeping your spine in a neutral position and your hips and knees slightly bent

DUMBBELL SHRUG

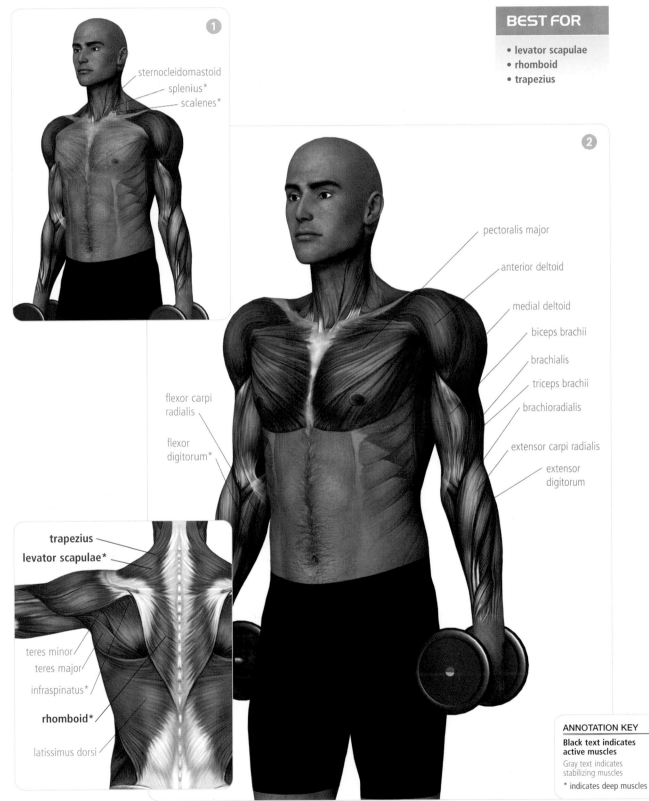

1

sternocleidomastoid
splenius*
scalenes*

BEST FOR

- **levator scapulae**
- **rhomboid**
- **trapezius**

2

pectoralis major

anterior deltoid

medial deltoid

biceps brachii

brachialis

triceps brachii

brachioradialis

extensor carpi radialis

extensor digitorum

flexor carpi radialis

flexor digitorum*

trapezius
levator scapulae*

teres minor
teres major
infraspinatus*

rhomboid*

latissimus dorsi

ANNOTATION KEY

Black text indicates active muscles

Gray text indicates stabilizing muscles

* indicates deep muscles

OVERHEAD PRESS

1 **Starting Position:** Seated on a Swiss ball, begin with your hands parallel with your collarbone and slightly wider than shoulder-width, your elbows pointing directly down at the floor, and your palms away from you. Keep your spine in a neutral position and look directly forward. Draw your abdomen in and up and keep your chest and ribs high.

LOOK FOR
- A smooth transition from the bottom to the top of the movement
- The identical speed and movement path with each hand

AVOID
- Extending your chin forward
- Elevating your shoulders
- Rounding your back
- Accelerating the weight by altering the speed with which you move toward the ceiling
- Any end position other than holding the weight directly over your shoulders

Action: Exhale, pushing your palms directly up toward the ceiling. Keep your shoulder blades down.

Movement Path: Your hands should remain in the same plane as your torso.

Your arms should finish with your hands directly above your shoulder joints and your shoulders down and away from your ears.

Arc slightly through the push, making sure your bone structure is supporting the weight.

2

STABILIZE BY
- Pulling your abdominal muscles up and in
- Maintaining a neutral spinal position
- Keeping your knees directly over your feet and your feet pressed firmly into the floor
- Keeping your shoulder blades down and your ribcage high

OVERHEAD PRESS

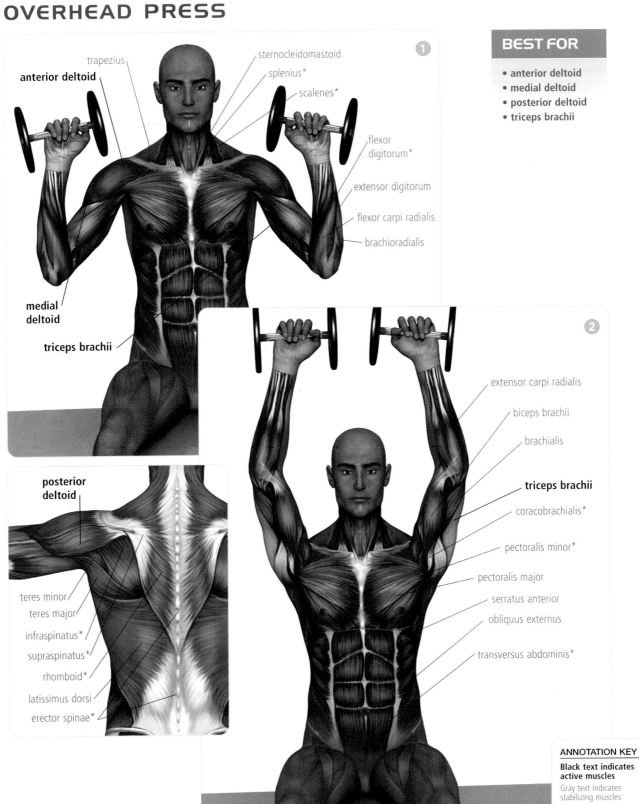

BEST FOR

- anterior deltoid
- medial deltoid
- posterior deltoid
- triceps brachii

1

trapezius
sternocleidomastoid
splenius*
anterior deltoid
scalenes*
flexor digitorum*
extensor digitorum
flexor carpi radialis
brachioradialis

medial deltoid

triceps brachii

2

extensor carpi radialis
biceps brachii
brachialis
triceps brachii
coracobrachialis*
pectoralis minor*
pectoralis major
serratus anterior
obliquus externus
transversus abdominis*

posterior deltoid

teres minor
teres major
infraspinatus*
supraspinatus*
rhomboid*
latissimus dorsi
erector spinae*

ANNOTATION KEY

**Black text indicates
active muscles**

Gray text indicates
stabilizing muscles

* indicates deep muscles

ARMS ANATOMY

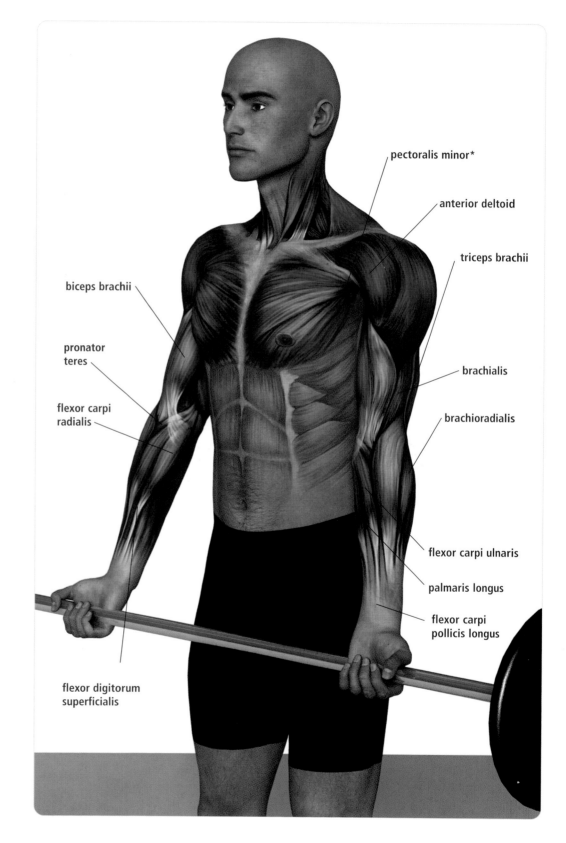

pectoralis minor*

anterior deltoid

triceps brachii

biceps brachii

brachialis

pronator teres

brachioradialis

flexor carpi radialis

flexor carpi ulnaris

palmaris longus

flexor carpi pollicis longus

flexor digitorum superficialis

ANNOTATION KEY

* indicates deep muscles

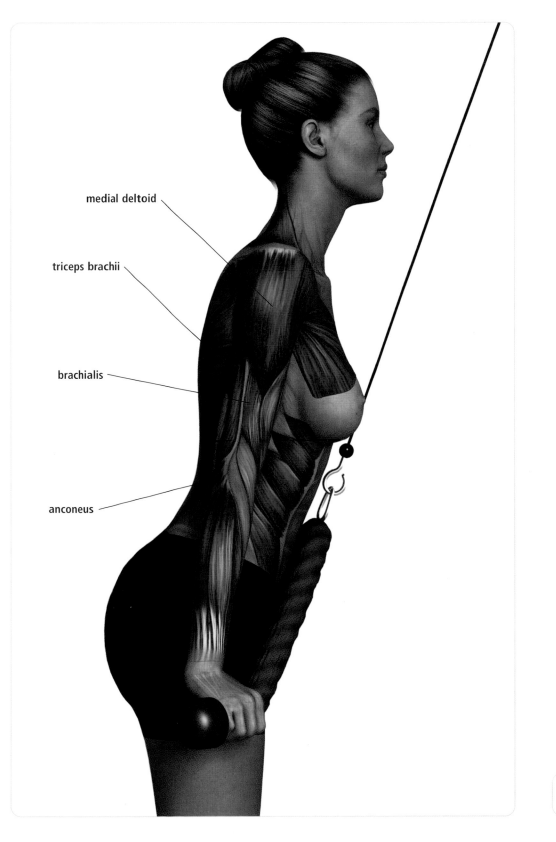

medial deltoid

triceps brachii

brachialis

anconeus

BICEP CURL

Starting Position: Stand straight up with your knees slightly bent. Keep your chest up and shoulders down. Grasp dumbbells with your palms facing your thighs, directly at your sides.

LOOK FOR
- Your elbows to stay close to your body
- Your head to stay upright and your shoulder blades to stay down
- Your palms to rotate gradually as the weights are lifted and lowered

AVOID
- Extending your back during the curl; your spine should remain neutral
- Elevating your shoulders
- Moving your elbows too far away from your body
- Moving your head forward
- Letting the dumbbells down in an uncontrolled manner during the eccentric phase
- Any torso rotation
- Swinging the weights up

Action: Exhale and bring the weights up, rotating your hands so that your thumbs turn outward as your hands rise, keeping your elbows tightly at your sides. Finish with your palms facing directly toward your shoulder joints. Inhale and return the weights in a slow, controlled manner.

Movement Path: Your torso, hips, and legs are motionless as your lower arms are drawn upward in an arc from a vertical position in which they are pointing down, to one in which your hands are adjacent to your shoulder joints.

STABILIZE BY
- Retracting your shoulder blades throughout the entire movement
- Pulling your abdomen up and in

BICEP CURL

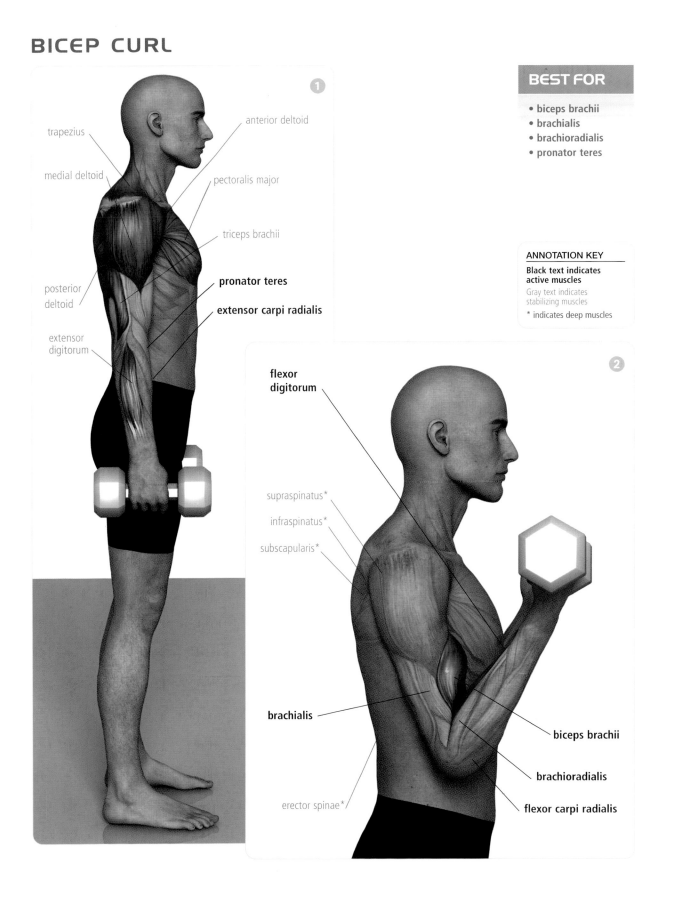

①

trapezius

anterior deltoid

medial deltoid

pectoralis major

triceps brachii

pronator teres

extensor carpi radialis

posterior deltoid

extensor digitorum

ANNOTATION KEY

Black text indicates active muscles
Gray text indicates stabilizing muscles
* indicates deep muscles

②

flexor digitorum

supraspinatus*

infraspinatus*

subscapularis*

brachialis

biceps brachii

brachioradialis

erector spinae*

flexor carpi radialis

BARBELL CURL

Starting Position: Grasp the barbell with your palms facing forward (away from your body) and with your hands slightly wider than your hips. Your torso should be upright and your hips and knees should be slightly bent. Keep your abdomen pulled up and in, your chest high, and your chin up.

LOOK FOR
- Your torso to remain stationary
- The bar to move at one rate through the full range of the motion
- Your shoulders to stay down, your chest to stay up, and your abdomen to stay pulled in

AVOID
- Swinging the bar
- Leaning backward
- Allowing your elbows to slide backward during the upward movement of the bar
- Moving your head forward

Action: Exhale and bring the weight up, keeping your elbows tightly at your sides. Finish with your palms facing directly backward. Inhale and return the barbell to the starting position in a slow, controlled manner.

Movement Path: Your torso, hips, and legs are motionless as your lower arms are drawn upward in an arc, from a vertical position in which they point down, to one in which your hands are adjacent to your shoulder joints.

STABILIZE BY
- Keeping your elbows in by your sides and pulled back, parallel with your spine
- Pulling your abdomen up and in
- Keeping your spine neutral and your shoulders down and back

BARBELL CURL

BEST FOR

- **brachioradialis**
- **biceps brachii**

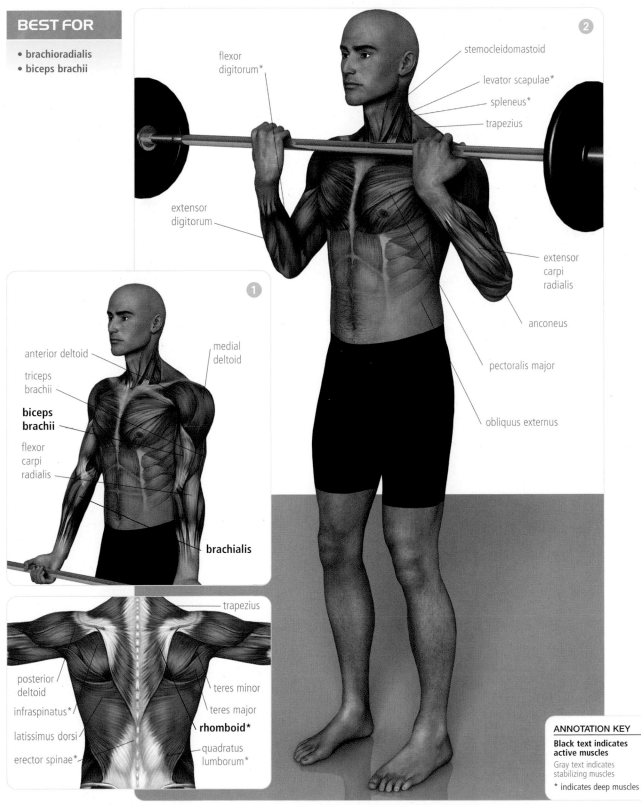

2

flexor digitorum*

stemocleidomastoid

levator scapulae*

spleneus*

trapezius

extensor digitorum

extensor carpi radialis

anconeus

pectoralis major

obliquus externus

1

anterior deltoid

triceps brachii

biceps brachii

flexor carpi radialis

medial deltoid

brachialis

trapezius

posterior deltoid

infraspinatus*

latissimus dorsi

erector spinae*

teres minor

teres major

rhomboid*

quadratus lumborum*

ANNOTATION KEY

Black text indicates active muscles

Gray text indicates stabilizing muscles

* indicates deep muscles

CORE

The muscles of the core originate on either the pelvis or spinal column and are responsible for forward flexion, extension, lateral flexion and rotation of the spine. These muscles are intimately connected to the muscles of the hip and back and are critical to the proper functioning of both.

Along with hip and back muscles, there are both superficial and deep layers, some of which include muscles (e.g., the transversus abdominis) that are responsible only for stabilization, not movement. Core training has typically consisted of movements that are primarily forward flexion and rotation biased; however, understanding that each plane of movement is necessary for proper function and, subsequently, injury prevention, is imperative.

All athletic activities require the core muscles to stabilize and translate forces from the lower to upper body and vice versa.

CORE ANATOMY

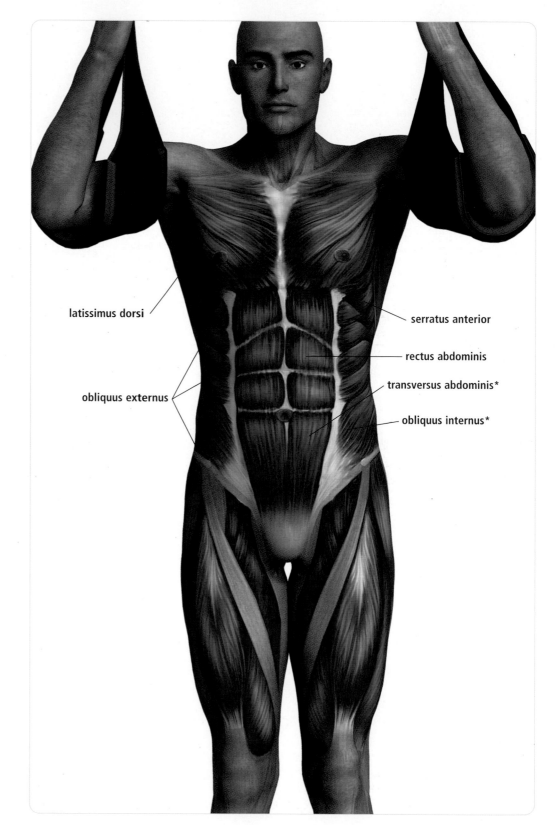

latissimus dorsi

serratus anterior

rectus abdominis

transversus abdominis*

obliquus externus

obliquus internus*

ANNOTATION KEY

* indicates deep muscles

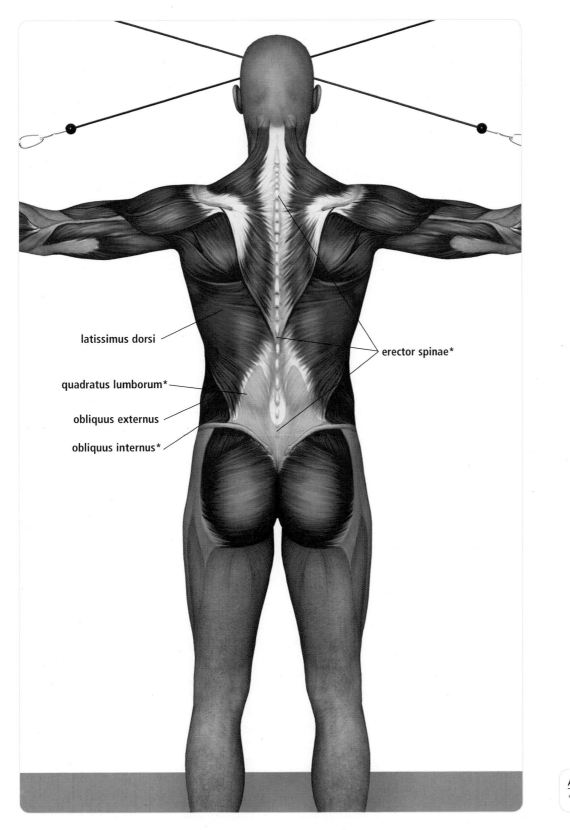

latissimus dorsi

quadratus lumborum*

obliquus externus

obliquus internus*

erector spinae*

ANNOTATION KEY

* indicates deep muscles

TRANSVERSE ABDOMINALS

Starting Position: Sit on a Swiss ball with your hands on your hips. Your shoulders should be down and your head should face directly forward.

LOOK FOR
- Your chest to remain high and your ribs to remain up and apart

AVOID
- Torso flexion (bending forward) or any hip movement

Action: Inhale through your lower abdomen, pushing your bellybutton out. Exhale and drag your bellybutton in and up, elevating your rib cage and cinching your waist.

Movement Path: Your abdomen contracts backward toward your spine.

STABILIZE BY
- Keeping your chest up and your shoulders down and relaxed
- Maintaining a neutral spinal position

TRANSVERSE ABDOMINALS

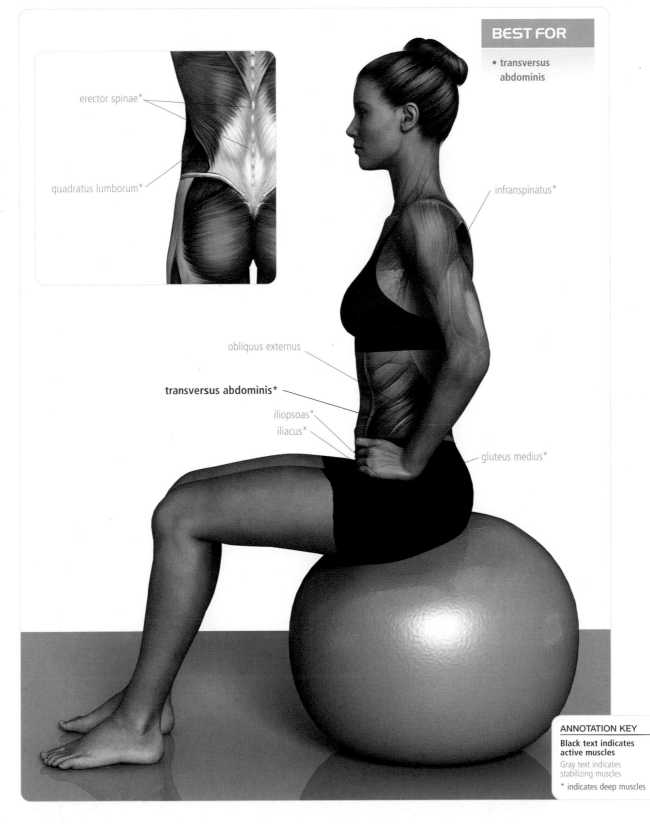

erector spinae*

quadratus lumborum*

BEST FOR

• transversus abdominis

infranspinatus*

obliquus externus

transversus abdominis*

iliopsoas*

iliacus*

gluteus medius*

ANNOTATION KEY

Black text indicates active muscles

Gray text indicates stabilizing muscles

* indicates deep muscles

CRUNCH

Starting Position: Place your hands behind your head while lying flat on the ground. Bend your knees so that your feet are flat on the floor, keeping your spine long. ❶

LOOK FOR
- A smooth movement throughout the entire length of your spine
- Your abdominal muscles to contract and pull in
- Your hips to remain stable

AVOID
- Pulling with your hands, bringing your chin toward your chest or collarbone, arching your back, or elevating your feet

Action: Push your lower back into the ground, keeping your spine long. Contract your abdominal muscles and lift your upper back off the ground and slightly forward, exhaling as you come up.

Movement Path: Your torso curves from your mid-low back to the top of your head, in straight line up and toward the knees. ❷

STABILIZE BY
- Keeping your shoulders down with your elbows widely spread
- Keeping your hips even and your feet flat

CRUNCH

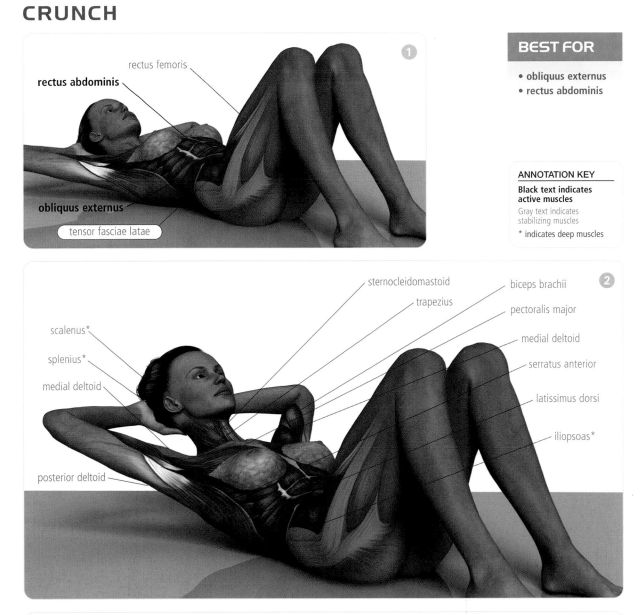

①
- rectus femoris
- **rectus abdominis**
- **obliquus externus**
- tensor fasciae latae

②
- scalenus*
- splenius*
- medial deltoid
- posterior deltoid
- sternocleidomastoid
- trapezius
- biceps brachii
- pectoralis major
- medial deltoid
- serratus anterior
- latissimus dorsi
- iliopsoas*

MODIFICATIONS

Similar Difficulty: Place your feet on a Swiss ball while you do a crunch. Do not let the ball roll.

Similar Difficulty: Do a regular crunch while sitting on a Swiss ball. Do not let the ball roll.

More Difficult: Raise your shins up to 45 degree angles.

More Difficult: Lift a medicine ball from your chest straight up. Keep your arms straight as you do a crunch.

More Difficult: As before, but raise one leg and extend the ball to your out-stretched foot as you do a crunch.

AB WHEEL

Starting Position: On your knees, bend your torso forward at a 45 degree angle with your spine in a neutral position. Extend your arms forward at a 45 to 90 degree angle to your torso, with your elbows straight and your hands grasping the wheel.

LOOK FOR
- All joints to move at the same time
- Your head and spine to remain aligned

AVOID
- Rounding or arching your spine
- Allowing your joints to move sequentially
- Moving quickly in either direction

Action: Inhale and extend your arms forward, allowing your torso to drop until your chest is almost parallel to the floor, rolling the wheel in a straight line away from you. Your hips move forward, following your torso, but your knees remain stationary. Exhale and draw your arms and hips back simultaneously; your torso elevates and returns to the starting position.

Movement Path: Your center of mass is translated forward and downward as your arms and hips extend into a linear position, with your knees as the fulcrum.

STABILIZE BY
- Pulling your abdomen up and in
- Keeping your shoulders down and back throughout the movement
- Keeping your arms extended and your wrists solid
- Maintaining a neutral spinal position throughout the movement

AB WHEEL

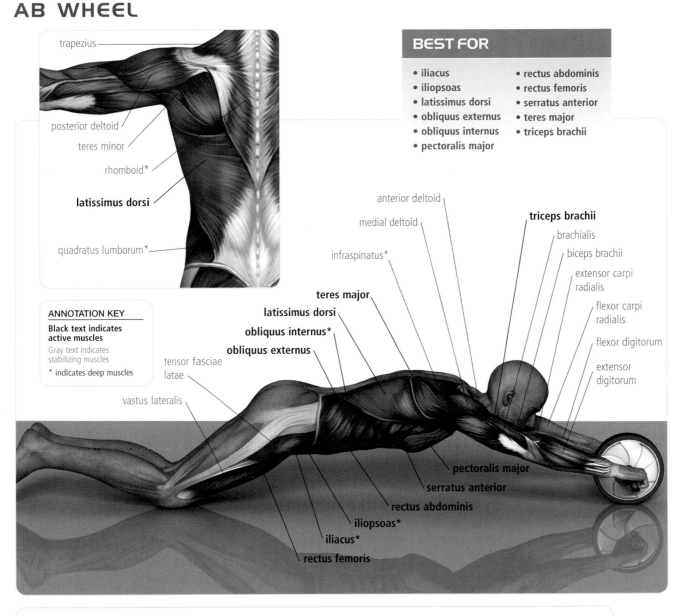

trapezius

posterior deltoid

teres minor

rhomboid*

latissimus dorsi

quadratus lumborum*

BEST FOR

- iliacus
- iliopsoas
- latissimus dorsi
- obliquus externus
- obliquus internus
- pectoralis major
- rectus abdominis
- rectus femoris
- serratus anterior
- teres major
- triceps brachii

anterior deltoid

medial deltoid

infraspinatus*

teres major

latissimus dorsi

obliquus internus*

obliquus externus

tensor fasciae latae

vastus lateralis

triceps brachii

brachialis

biceps brachii

extensor carpi radialis

flexor carpi radialis

flexor digitorum

extensor digitorum

pectoralis major

serratus anterior

rectus abdominis

iliopsoas*

iliacus*

rectus femoris

ANNOTATION KEY

Black text indicates active muscles

Gray text indicates stabilizing muscles

* indicates deep muscles

MODIFICATIONS

Similar Difficulty: Replace the wheel with a Swiss ball; your hands begin higher up.

1

2

HANGING LEG RAISE

STABILIZE BY
- Keeping your upper arms parallel and your shoulders down
- Gripping the stirrups firmly
- Keeping your legs parallel

LOOK FOR
- Your knees to bend as your upper legs are raised (your lower legs remain vertical)
- Your legs to move upward together

AVOID
- Swinging
- Extending your arms upward to more than 5 degrees above horizontal
- Moving your hips backward

Starting Position: Hang with your upper arms in stirrups, with your elbows bent at 90 degree angles, pointing forward just above shoulder height. Grasp the stirrups with your hands and make sure your torso, legs, and hips are straight.

Action: Pull your upper arms and elbows downward and your upper legs and knees upward toward your elbows, flexing at the hips. Tuck your hips forward and bring your chest forward slightly. Return to the starting position in a slow and controlled manner. Exhale as you rise and inhale as you return to the starting position.

Movement Path: You torso rounds slightly as your hips flex upward and your upper arms are pulled downward. Your center of mass makes no appreciable movement.

HANGING LEG RAISE

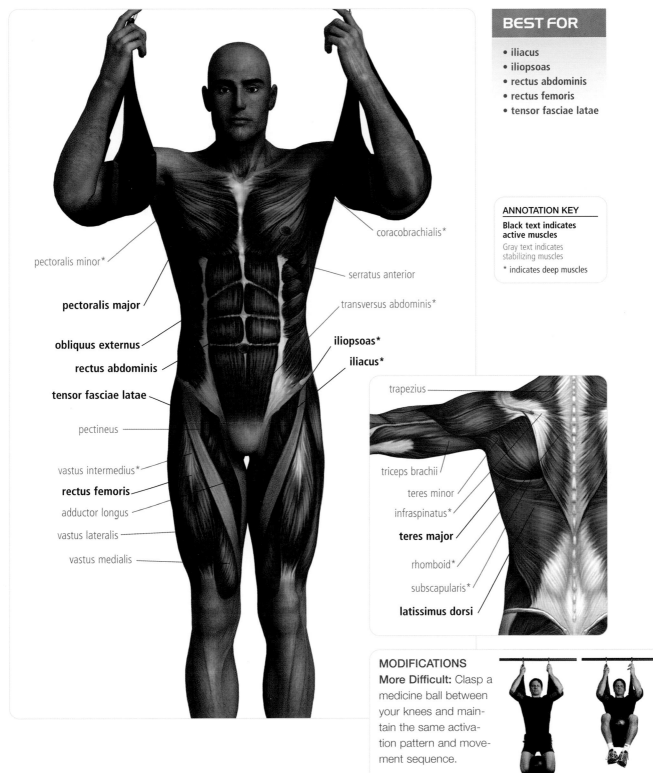

pectoralis minor*

pectoralis major

obliquus externus

rectus abdominis

tensor fasciae latae

pectineus

vastus intermedius*

rectus femoris

adductor longus

vastus lateralis

vastus medialis

coracobrachialis*

serratus anterior

transversus abdominis*

iliopsoas*

iliacus*

trapezius

triceps brachii

teres minor

infraspinatus*

teres major

rhomboid*

subscapularis*

latissimus dorsi

BEST FOR

- iliacus
- iliopsoas
- rectus abdominis
- rectus femoris
- tensor fasciae latae

ANNOTATION KEY

Black text indicates active muscles

Gray text indicates stabilizing muscles

* indicates deep muscles

MODIFICATIONS

More Difficult: Clasp a medicine ball between your knees and maintain the same activation pattern and movement sequence.

① ②

BRIDGE

Starting Position: Bend your knees, with your feet flat on the floor. Bend your elbows to 90 degree angles, with your hands facing the ceiling and a slight arch in your lower back, which should not contact the ground.

1

LOOK FOR
- Your hip to be the hinge; there should be no movement in your spine from the hip to the shoulder joint

AVOID
- Sequential lifting, a pelvic tuck, or any rotational elevation (one hip rising faster than the other)

Action: Using your upper arms, shoulders, and feet, push your hips and ribs simultaneously up toward the ceiling. Return to the starting position.

Movement Path: Curvilinear; the spinal movement is straight up from the floor.

2

STABILIZE BY
- Keeping your upper arms and elbows pulled down and into the floor
- Distributing your weight evenly, keeping your feet flat
- Keeping your hips, knees, and feet in a single line

BRIDGE

BEST FOR

- biceps femoris
- gluteus maximus
- rhomboid
- quadratus lumborum
- semimembranosus
- semitendinosus

ANNOTATION KEY

Black text indicates active muscles

Gray text indicates stabilizing muscles

* indicates deep muscles

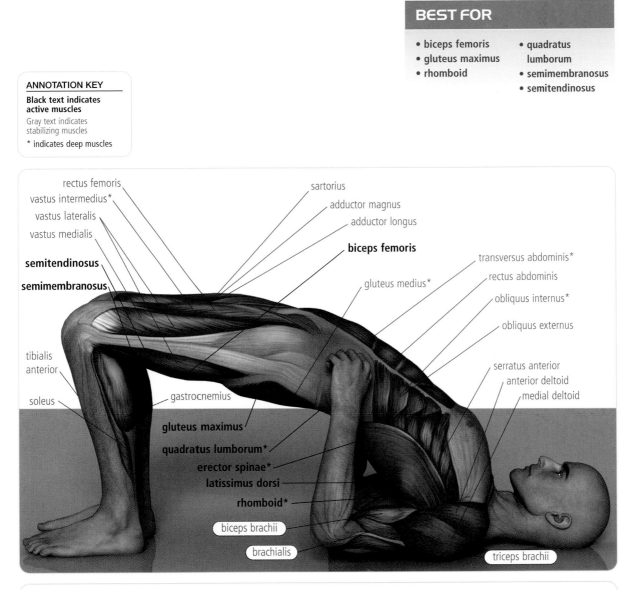

rectus femoris
vastus intermedius*
vastus lateralis
vastus medialis
semitendinosus
semimembranosus
tibialis anterior
soleus
gastrocnemius
gluteus maximus
quadratus lumborum*
erector spinae*
latissimus dorsi
rhomboid*
biceps brachii
brachialis
sartorius
adductor magnus
adductor longus
biceps femoris
gluteus medius*
transversus abdominis*
rectus abdominis
obliquus internus*
obliquus externus
serratus anterior
anterior deltoid
medial deltoid
triceps brachii

MODIFICATIONS

More Difficult: Raise one leg and maintain the same activation pattern and movement sequence.

More Difficult: Lay your arms flat and balance your feet on a Swiss ball, keeping your hips elevated and the plane of your body flat.

More Difficult: Same as previous, but bend one leg toward your body.

More Difficult: With your legs on a Swiss ball, raise one leg straight up in the top position.

ARM-LEG EXTENSION

Starting Position: Lie flat on the ground, with one arm bent, your elbow on the ground, your palm down, and the hand under your chin. Extend your other arm, holding your thumb up toward the ceiling.

Action: Simultaneously lift your extended arm, torso, and the opposite leg.

Movement Path: Your arm and leg rise straight up to form an arc with your torso.

LOOK FOR
- A simultaneous movement of the raised arm and leg
- An equivalent range or distance from the ground for the raised arm and leg
- Your hip bones to remain in contact with the ground

AVOID
- Any rotation of your torso or hips from the ground
- Elevating your shoulder or bending your knee or elbow

STABILIZE BY
- Keeping your shoulder blades down and back
- Keeping your hips even and down
- Keeping your legs and arms straight

MODIFICATIONS

More Difficult: Raise both arms and legs simultaneously.

More Difficult: On your hands and knees, place a medicine ball on your back. Extend one arm and the opposite leg (stabilizers: all spinal and hip musculature).

More Difficult: In a front plank position, raise one arm and the opposite leg (stabilize with all core muscles).

ARM-LEG EXTENSION

BEST FOR

- erector spinae
- gluteus maximus
- infraspinatus
- quadratus lumborum
- rhomboid
- splenius
- teres minor
- trapezius

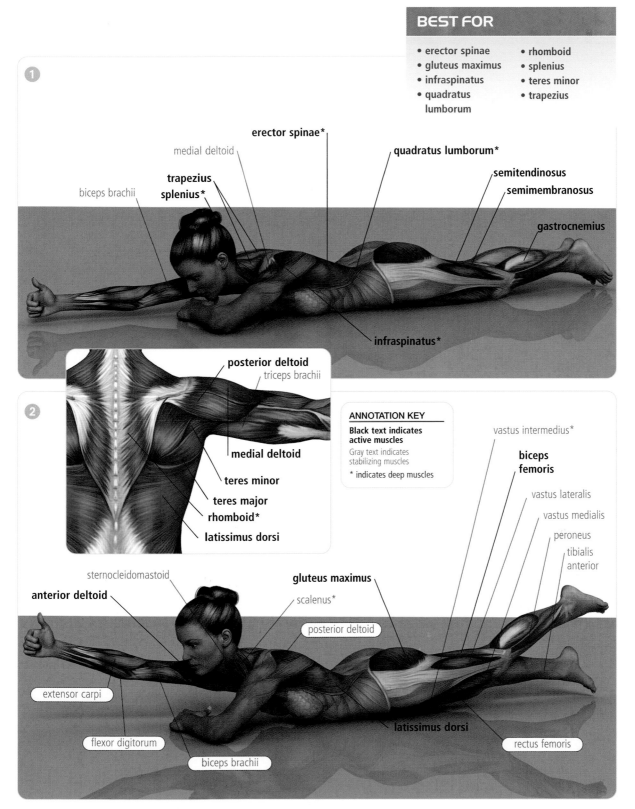

1

erector spinae*

medial deltoid

trapezius

splenius*

biceps brachii

quadratus lumborum*

semitendinosus

semimembranosus

gastrocnemius

infraspinatus*

posterior deltoid

triceps brachii

2

medial deltoid

teres minor

teres major

rhomboid*

latissimus dorsi

ANNOTATION KEY

Black text indicates active muscles

Gray text indicates stabilizing muscles

* indicates deep muscles

vastus intermedius*

biceps femoris

vastus lateralis

vastus medialis

peroneus

tibialis anterior

sternocleidomastoid

anterior deltoid

gluteus maximus

scalenus*

posterior deltoid

extensor carpi

flexor digitorum

biceps brachii

latissimus dorsi

rectus femoris

FRONT PLANK

Starting Position: Lie face down on the ground and fold your hands directly beneath your chin with your elbows by your sides and your feet on your toes.

Action: Raise the length of your torso to a horizontal position with a slight arch in your lower back. Your shoulder blades should be flat and your spine long.

Movement Path: None.

LOOK FOR
- A neutral spinal position
- Locked knees, with your ankles at 90 degree angles and your elbows directly under your shoulder joints

AVOID
- Rounding your spine, dropping your hips, and elevating your shoulders toward your ears

STABILIZE BY
- Keeping your spine neutral
- Keeping your shoulders down and your head up
- Maintaining the contraction of your gluteals and legs
- Keeping your legs straight and your ankles bent at 90 degree angles, with your toes pointing directly into the ground

MODIFICATIONS
Easier: Raise your forelegs and rest your weight on your knees to shorten the lever.

FRONT PLANK

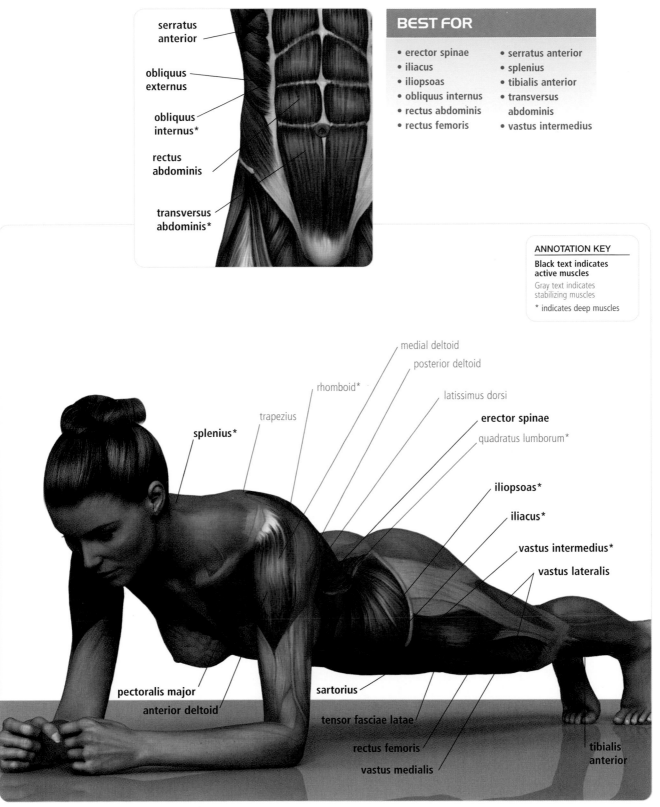

serratus anterior

obliquus externus

obliquus internus*

rectus abdominis

transversus abdominis*

BEST FOR

- erector spinae
- iliacus
- iliopsoas
- obliquus internus
- rectus abdominis
- rectus femoris
- serratus anterior
- splenius
- tibialis anterior
- transversus abdominis
- vastus intermedius

ANNOTATION KEY

Black text indicates active muscles

Gray text indicates stabilizing muscles

* indicates deep muscles

medial deltoid

posterior deltoid

rhomboid*

latissimus dorsi

trapezius

erector spinae

quadratus lumborum*

splenius*

iliopsoas*

iliacus*

vastus intermedius*

vastus lateralis

pectoralis major

anterior deltoid

sartorius

tensor fasciae latae

rectus femoris

vastus medialis

tibialis anterior

SIDE PLANK

Starting Position: Lie on your side with the lower arm bent at the elbow. The lower elbow should be underneath the shoulder joint and the upper hand should be on your hip. Align your ankles, hips, shoulders, and head.

Action: Push the length of your body toward the ceiling, balancing on the edge of your bottom shoe with one foot directly over the other.

Movement Path: Your center of mass rises directly upward.

LOOK FOR
- Your spine to be in a straight position, similar to standing
- Your elbow to be directly under your shoulder joint, with a slight arch in your lower back, your knees straight, and your ankles at 90 degree angles

AVOID
- Rotating your hips, allowing your shoulder blades to slide toward your spine, or moving your head forward

STABILIZE BY
- Pulling your abdomen up and in, keeping your spine neutral
- Keeping your shoulder blades away from your spine and flat on your ribcage
- Straightening and contracting your legs
- Balancing on the edge of your bottom foot or shoe

SIDE PLANK

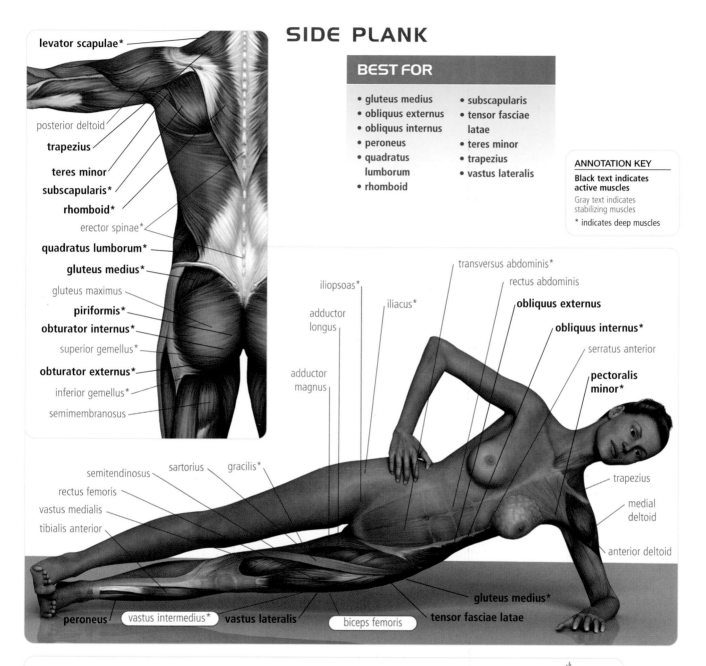

BEST FOR

- gluteus medius
- obliquus externus
- obliquus internus
- peroneus
- quadratus lumborum
- rhomboid
- subscapularis
- tensor fasciae latae
- teres minor
- trapezius
- vastus lateralis

ANNOTATION KEY
Black text indicates active muscles
Gray text indicates stabilizing muscles
* indicates deep muscles

levator scapulae*
posterior deltoid
trapezius
teres minor
subscapularis*
rhomboid*
erector spinae*
quadratus lumborum*
gluteus medius*
gluteus maximus
piriformis*
obturator internus*
superior gemellus*
obturator externus*
inferior gemellus*
semimembranosus

iliopsoas*
adductor longus
adductor magnus
iliacus*
transversus abdominis*
rectus abdominis
obliquus externus
obliquus internus*
serratus anterior
pectoralis minor*
trapezius
medial deltoid
anterior deltoid

semitendinosus
sartorius
gracilis*
rectus femoris
vastus medialis
tibialis anterior
peroneus
vastus intermedius*
vastus lateralis
biceps femoris
tensor fasciae latae
gluteus medius*

MODIFICATIONS

Easier: Bend your lower leg so that, keeping your knees parallel, your weight is balanced on your bent knee.

Easier: Bend both legs, crossing your top leg in front so that both knees are on the ground. Push

your hip and torso toward the ceiling, extending the top leg and knee up and out.

More Difficult: Raise your top arm and balance with both arms straightened.

SANTANA PUSH-UP

Starting Position: Lie flat on the floor with your hands slightly wider than shoulder width, grasping dumbbells, so that the dumbbell handles are parallel to your spine. Point your elbows directly at the ceiling. Your feet should be slightly wider than shoulder width and your spine should be neutral.

①

Action: Push up toward the ceiling; once your arms are fully extended, rotate your hips and feet, lifting one arm in an arc toward the ceiling so that your arms are aligned in a straight line and your feet are split apart, with your weight on the edges of your shoes. Your torso, hips, and legs are rigid.

②

LOOK FOR
- Your shoulders to remain depressed, your neck to remain long, your hips to remain elevated, and your shoulders, hips, and feet to remain in the same plane from the floor

AVOID
- Bending your knees, dropping your hips, or excessive rotation in shoulder and hip

Movement Path: Your entire body moves up and away from the floor, and then rotates around your spine 180 degrees.

③

STABILIZE BY
- Pulling your abdomen up and in
- Keeping your shoulder blades down and flat
- Keeping your knees straight and your legs contracted
- Maintaining a neutral spinal position throughout the movement

SANTANA PUSH-UP

1

subscapularis*
posterior deltoid
supraspinatus*
triceps brachii
trapezius
splenius*
teres minor
teres major
obliquus externus
obliquus internus*
obturator externus*
obturator internus*
adductor magnus
gastrocnemii
peroneii
sternocleidomastoid
medial deltoid
tensor fascia latae

BEST FOR

- adductor longus
- adductor magnus
- anterior deltoid
- coracobrachialis
- gluteus medius
- gracilis
- obturator externus
- obturator internus
- pectoralis major
- piriformis
- posterior deltoid
- quadratus lumborum
- sartorius
- subscapularis
- supraspinatus
- tensor fasciae latae
- teres major
- teres minor
- vastus lateralis

2

latissimus dorsi
triceps brachii
erector spinae
quadratus lumborum*
gluteus medius*
piriformis*
iliacus*
iliopsoas*

3

anterior deltoid
sternocleidomastoid
levator scapulae*
biceps brachii
serratus anterior
pectoralis major
scalenes*
biceps brachii
extensor carpi
coracobrachialis*
tensor fasciae latae
rectus abdominis
transversus abdominis*
vastus intermedius
gracilis
pectineus*
sartorius
adductor longus
vastus lateralis
vastus medialis
soleus
tibialis anterior
flexor carpi
flexor digitorum
extensor digitorum

ANNOTATION KEY

Black text indicates active muscles
Gray text indicates stabilizing muscles
* indicates deep muscles

PNF RAISE

①

STABILIZE BY
- Pulling your abdomen up and in
- Distributing your weight evenly across your foot
- Using all muscles and joints in a coordinated, relaxed manner

Starting Position: Stand on one foot, bending the raised knee, and grasp a medicine ball just below and to the outside of the knee on the standing leg.

②

LOOK FOR
- Your knee and hip to extend and rise at the same time
- The ball to remain equidistant from your torso throughout the movement
- Your elbows to remain extended

AVOID
- Excessive flexion of your torso and spine
- Bringing the ball close to your body or lifting any part of your foot from the floor

Action: Stand, extending your leg, while bringing the ball across your body to above and outside the opposite shoulder.

Movement Path: Your upper body rotates as your center of mass shifts upward. The ball moves in an arc across your body.

PNF RAISE

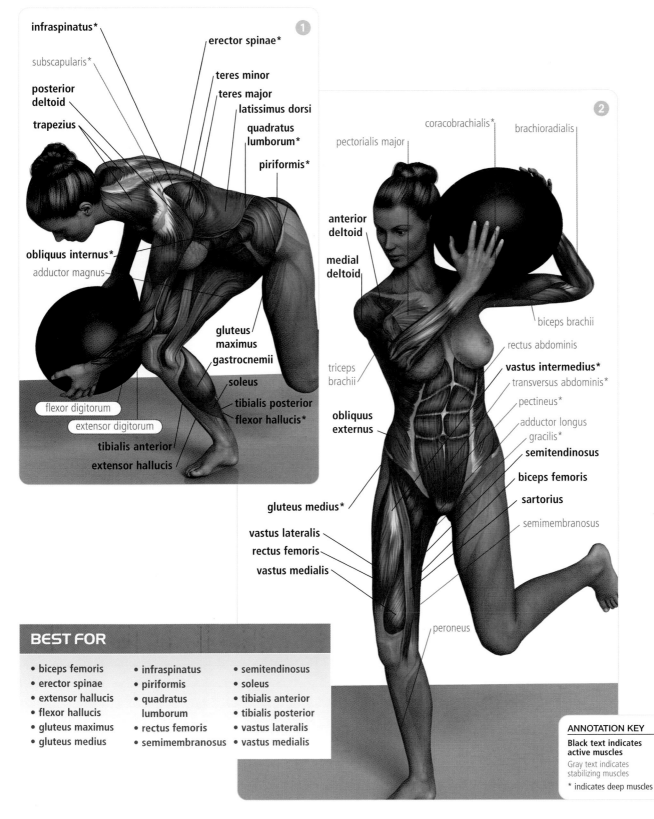

1

infraspinatus*

erector spinae*

subscapularis*

teres minor

posterior
deltoid

teres major

latissimus dorsi

trapezius

quadratus
lumborum*

piriformis*

obliquus internus*

adductor magnus

gluteus
maximus

gastrocnemii

soleus

flexor digitorum

tibialis posterior

extensor digitorum

flexor hallucis*

tibialis anterior

extensor hallucis

2

coracobrachialis*

brachioradialis

pectorialis major

anterior
deltoid

medial
deltoid

biceps brachii

rectus abdominis

triceps
brachii

vastus intermedius*

transversus abdominis*

pectineus*

obliquus
externus

adductor longus

gracilis*

semitendinosus

biceps femoris

sartorius

semimembranosus

gluteus medius*

vastus lateralis

rectus femoris

vastus medialis

peroneus

BEST FOR

- biceps femoris
- erector spinae
- extensor hallucis
- flexor hallucis
- gluteus maximus
- gluteus medius
- infraspinatus
- piriformis
- quadratus
  lumborum
- rectus femoris
- semimembranosus
- semitendinosus
- soleus
- tibialis anterior
- tibialis posterior
- vastus lateralis
- vastus medialis

ANNOTATION KEY

**Black text indicates
active muscles**

Gray text indicates
stabilizing muscles

* indicates deep muscles

WOODCHOPPER

1

Starting Position: Stand with the cable pulley to one side and slightly forward of the plane of your body, with your feet wider than your shoulders. Grasp the handle with both hands at shoulder height by crossing your torso with the opposite arm. Align your shoulder, hips, and ankles in the same plane.

2

Action: Pull the handle downward and inward with straight arms in a 90 degree arc, stopping in front of the opposite leg; return along the same path. Exhale as you pull the weight down and inhale as you return to the starting position.

Movement Path: Your torso and center of mass remain stationary as your arms move in a 90 degree arc from the side of your body downward and inward until your hands cross your midline.

LOOK FOR
- Your torso and hips to remain stationary
- Your arms to be fully extended at the bottom of the movement
- Elevating your chest at the bottom of the movement

AVOID
- Rotating your hips
- Bending the elbow of the arm crossing your body
- Shifting your weight from side to side during the movement

STABILIZE BY
- Keeping your shoulders down and back
- Pulling your abdomen up and in and keeping your chest high
- Distributing your weight evenly across your feet
- Keeping your hips and knees slightly bent and solid

WOODCHOPPER

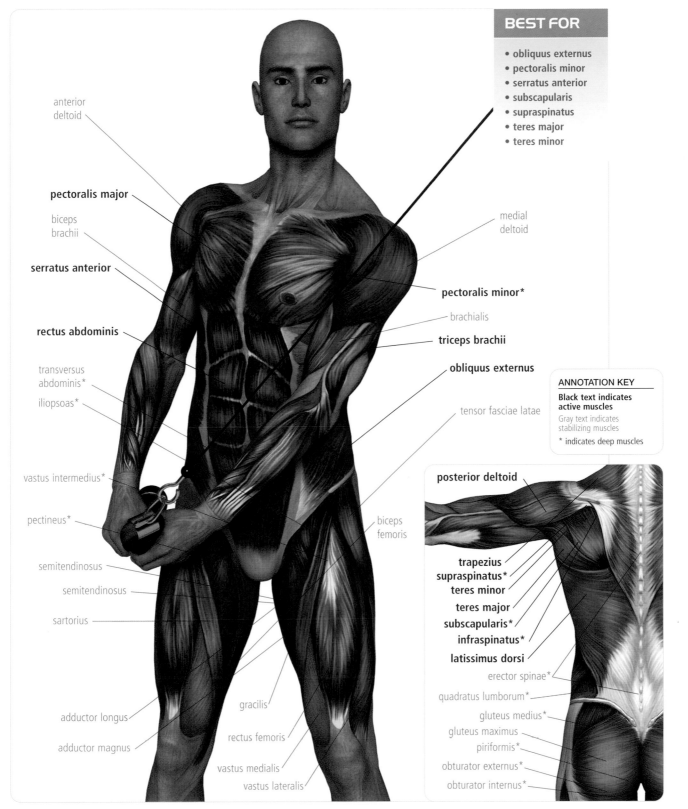

BEST FOR

- obliquus externus
- pectoralis minor
- serratus anterior
- subscapularis
- supraspinatus
- teres major
- teres minor

anterior deltoid

pectoralis major

biceps brachii

serratus anterior

rectus abdominis

transversus abdominis*

iliopsoas*

vastus intermedius*

pectineus*

semitendinosus

semitendinosus

sartorius

adductor longus

adductor magnus

medial deltoid

pectoralis minor*

brachialis

triceps brachii

obliquus externus

tensor fasciae latae

biceps femoris

gracilis

rectus femoris

vastus medialis

vastus lateralis

ANNOTATION KEY

Black text indicates active muscles

Gray text indicates stabilizing muscles

* indicates deep muscles

posterior deltoid

trapezius
supraspinatus*
teres minor
teres major
subscapularis*
infraspinatus*
latissimus dorsi

erector spinae*

quadratus lumborum*

gluteus medius*

gluteus maximus

piriformis*

obturator externus*

obturator internus*

INDEX

This checklist provides a quick and easy way to keep track of exercises. Once you have mastered the techniques described for these exercises and their variations, you can use this list to devise different workouts to keep your exercise sessions fresh and productive.

Lateral Lunge
p. 42

Step-Up
p. 44

Step-Down
p. 46

Calf Raise
p. 48

Cable Abduction
p. 50

Clamshells
p. 52

Single Leg Deadlift
p. 54

Leg Press Pliée
p. 56

Leg Extension
p. 58

Leg Curl
p. 60

Wall Sit
p. 62

Skater
p. 64

Plough
p. 66

Back Extension
p. 72

Romanian Deadlift
p. 74

One Arm Dumbbell Row
p. 76

Barbell Row
p. 78

Deadlift
p. 80

Chin-Up
p. 82

Body Row
p. 84

INDEX

Triceps Extension
p. 138

Bench Dip
p. 140

Bicep Curl
p. 142

Hammer Curl
p. 144

Wrist Curl
p. 146

Dumbbell Kickback
p. 148

Triceps Pushdown
p. 150

Barbell Curl
p. 152

Transverse Abdominals
p. 158

Crunch
p. 160

Cross-Over Crunch
p. 162

Obliques
p. 164

Ab Wheel
p. 166

Hanging Leg Raise
p. 168

Bridge
p. 170

Arm-Leg Extension
p. 172

Front Plank
p. 174

Side Plank
p. 176

Santana Push-Up
p. 178

PNF Raise
p. 180

Woodchopper
p. 182

GLOSSARY

Abduction: movement away from the body

Adduction: movement toward the body

Alternating Grip: one hand grasping with the palm facing toward the body and the other facing away

Anterior: located in the front

Extension: the act of straightening

Curvilinear (movement path): moving in a curved path

Flexion: the bending of a joint

Isometric: muscles contracting against an equal resistance, resulting in no movement

ITB: iliotibial band (see page 13 and 28)

Neutral Position (spine): a spinal position resembling an "S" shape, consisting of a *lordosis* in the lower back, when viewed in profile

PNF: Proprioceptive Neuromuscular Facilitation; refers to a neuromuscular pattern of contraction that uses the greatest efficiency regarding positional awareness

Posterior: located behind

Medial: located on, or extending toward, the middle

Lateral: located on, or extending toward, the outside

Lordosis: forward curvature of the spine and lumbar region

Scapula: the protrusion of bone on the mid-to upper-back, also known as the shoulder blade

Static: no movement; holding a given position

Dynamic: continuously moving

Active (also Active Isolated Stretching): Actively contracting a given muscle group (agonist) in order to stretch the opposing (antagonist) muscle group. Developed by Aaron Mattes.

Pose Technique: Running methodology developed by Dr. Nicholas Romanov.

LATIN GLOSSARY

The following glossary explains the Latin terminology used to describe the body's musculature. Certain words are derived from Greek, which has been indicated in each instance.

UPPER LEG

vastus lateralis: *vastus*, "immense, huge" and *lateralis*, "of the side"

vastus medialis: *vastus*, "immense, huge" and *medialis*, "middle"

vastus intermedius: *vastus*, "immense, huge" and *intermedius*, "that which is between"

rectus femoris: adj. of *rego*, "straight, upright" and *femur*, "thigh"

adductor longus: from *adducere*, "to contract" and *longus*, "long"

adductor magnus: from *adducere*, "to contract" and *magnus*, "major"

gracilis: *gracilis*, "slim, slender"

tensor fasciae latae: from *tenere*, "to stretch," *fasciae*, "band," and latae "laid down"

biceps femoris: *biceps*, "two-headed" and *femur*, "thigh"

semitendinosus: *semi*, "half" and *tendo*, "tendon"

semimembranosus: *semi*, "half" and *membrum*, "limb"

sartorius: from *sarcio*, "to patch" or "to repair"

LOWER LEG

gastrocnemii: Greek *gastroknémía*, "calf [of the leg]" and Latin suffix

soleus: *solea*, "sandal"

tibialis posterior: *tibia*, lit. "reed pipe" and *posterus*, "coming after"

tibialis anterior: *tibia*, lit. "reed pipe" and *ante*, "before"

peroneii: *peronei*, lit. "of the fibula"

flexor hallucis: from *flectere*, "to bend" and *hallex*, "big toe"

extensor hallucis: *extendere*, "to bend" and *hallex*, "big toe"

HIPS

gluteus medius: Greek *gloutós*, "rump," with Latin suffix, and *medialis*, "middle"

LATIN GLOSSARY

gluteus maximus: Greek *gloutós*, "rump," with Latin suffix, and *maximus*, "largest"

gluteus minimus: Greek *gloutós*, "rump," with Latin suffix, and *minimus*, "smallest"

iliopsoas: *ilia*, variant of *ilium*, "groin" and Greek *psoa*, "groin muscle"

iliacus: *ilia*, variant of *ilium*, "groin"

obturator externus: from *obturare*, "to block" and *externus*, "outward"

obturator internus: from *obturare*, "to block" and *internus*, "within"

pectineus: *pectin*, "comb"

superior gemellus: adj. comp. of *super*, "above" and *geminus*, "twin"

inferior gemellus: adj. of *inferus*, "under" and *geminus*, "twin"

piriformis: adj. of *pirum*, "pear"; therefore "pear-shaped"

quadratus femoris: *quadratus*, "square, rectangular" and *femur*, "thigh"

CORE

transversus abdominis: *transversus*, "athwart" and *abdomen*, "belly"

rectus abdominis: adj. of *rego*, "straight, upright" and *abdomen*, "belly"

obliquus internus: *obliquus*, "slanting" and *internus*, "within"

obliquus externus: *obliquus*, "slanting" and *externus*, "outward"

serratus anterior: adj. of *serra*, "saw"; therefore "saw-shaped" and *ante*, "before"

BACK

trapezius: Greek *trapezion*, lit. "small table"

rhomboid: Greek *rhembesthai*, "to spin"

latissimus dorsi: *latus*, "wide" and *dorsum*, "back"

erector spinae: *erectus*, "straight" and *spina*, "thorn"

quadratus lumborum: *quadratus*, "square, rectangular" and *lumbus*, "loin"

CHEST

pectoralis [major and minor]: *pectus*, "breast"

coracobrachialis: Greek *korakoeidés*, "raven-like" and *brachium*, "arm"

SHOULDERS

deltoid [anterior, posterior, and medial]: Greek *deltoeidés*, "delta-shaped"

supraspinatus: *supra*, "above" and *spina*, "thorn"

infraspinatus: *infra*, "under" and *spina*, "thorn"

subscapularis: *sub*, "below" and *scapulae*, "shoulder [blades]"

teres [major and minor]: *teres*, "rounded"

levator scapulae: from *levare*, "to raise" and *scapulae*, "shoulder [blades]"

UPPER ARM

biceps brachii: *biceps*, "two-headed" and *brachium*, "arm"

triceps brachii: *triceps*, "three-headed" and *brachium*, "arm"

brachialis: *brachium*, "arm"

LOWER ARM

brachioradialis: *brachium*, "arm" and *radius*, "spoke"

extensor carpi radialis: from *extendere*, "to bend", Greek *karpós*, "wrist", and *radius*, "spoke"

flexor carpi radialis: from *flectere*, "to bend", Greek *karpós*, "wrist", and *radius*, "spoke"

extensor digitorum: from *extendere*, "to bend" and *digitus*, "finger, toe"

flexor digitorum: from *flectere*, "to bend" and *digitus*, "finger, toe"

NECK

sternocleidomastoid: Greek *stérnon*, "chest", Greek *kleís*, "key", and Greek *mastoeidés*, "breast-like"

scalenes: Greek *skalénós*, "unequal"

splenius: Greek *splénion*, "plaster, patch"

ACKNOWLEDGMENTS

In my daily life for the past 20 years, several times a day I have been instructing people using the exercises in this book. I have worked with clients as young as 12 and as old as 92, from every different walk of life imaginable. From recreational athletes to professional ones, cardiac patients to pregnant and post-partum women, for entertainment or business, regardless of where they came from two things have remained constant for me: I have enjoyed every minute of it and felt privileged to be able to do it, and I am constantly reminded that exercise is relevant to everyone. Period.

As with any book, there are many people who have contributed to it. Their contributions also come in many different forms. For their help I would like to personally thank Sean, Karen, and Amber at Hylas Publishing for originally bringing this project to me and seeing it through. All of my staff here at La Palestra, Teodoro Chavez, Greg Cimino, Elisama Colon, Alex Evans, Gloria Robles, Shirley Gauthiers, Luis Gonzalez, Engenio Guerrero, Carrie Lane, Adam Palmer, Thierno Diallo, Jermaine Phanord, Ann Reginald, Molly Morgan, Sebaj Adele, Kofi Sekyiamah, Paul Knapic, Craig Maltese, Doug Dickinson, Melissa Mora, and Grace Eichinger deserve mention; and particularly for the contributions in helping with the text, Gillian Mounsey, Greg Peters, Marissa O'Neil, Garth Wakeford, and Sydney Foster, who along with Mark Tenore so skillfully and flawlessly modeled all of the exercises in this book. Dr. Rob DeStefano, Dr. Jennifer Solomon, and Dr. Kyler Brown were invaluable with their anatomical expertise, as was my assistant Shannon Plumstead who typed this entire work and is masterful at reading my handwriting. Thanks go to my good friends and colleagues Jim and Phil Wharton, who introduced me to the Technique of Active Isolated Stretching, and Nicholas Romanov, whose "Pose" method of running has had a profound influence on my thinking about all movement.

Lastly and most importantly I thank my family: my wife Deborah, who is my advisor on all things and whose opinion is the one I count on most when making any decision, my daughter Milena, and my son Tony (who was born while writing this), who make everything I do have a reason, and for my mom, who I thank every day for giving me the life that I have.